Praise for the Authors' Previous Title, *Galileo's New Universe*

"A charming peek into astronomy's 'family album,' this lively history is ideal for armchair scientists and stargazers."

—*Publishers Weekly*

"As a confirmed Galileo groupie, I applaud the way Maran and Marschall have taken the long view of our hero's achievements. . . . Their book makes the perfect link between telescopes then and now."

—Dava Sobel, author of *Galileo's Daughter*

"If you want a clear and lucid explanation of how over the last 400 years the telescope has transformed our view of the Moon, the Sun, planets, stars, and even our understanding of the structure of the Universe—read this book."

—Dr. Matt Mountain, director of the
Space Telescope Science Institute

"*Galileo's New Universe: The Revolution in our Understanding of the Cosmos* provides a deft two-step tour of the Galilean heavens—first as they were dimly glimpsed through the minds and instruments of 400 years ago and, second, in flashes forward to today's dramatically enhanced understanding, from the 27 million-degree nuclear-powered center of the sun to the 'flying saucer moons' hiding in Saturn's rings and beyond. Written in lively, jargon-free prose laced with wit and telling detail, the book is an entertaining anniversary tribute to all those who have used the telescope to 'see,' rather than merely look, into the depths of the Universe that spawned us all."

—Kathy Sawyer, author of *The Rock From Mars*

Pluto

CONFIDENTIAL

Pluto

CONFIDENTIAL

An Insider Account of the
Ongoing Battles over the
Status of Pluto

LAURENCE A. MARSCHALL
AND STEPHEN P. MARAN

BENBELLA

BENBELLA BOOKS, INC.
Dallas, TX

BENBELLA

BenBella Books, Inc.
6440 N. Central Expressway, Suite 503
Dallas, TX 75206
www.benbellabooks.com
Send feedback to feedback@benbellabooks.com

Printed in the United States of America
10 9 8 7 6 5 4 3 2 1

Library of Congress Cataloging-in-Publication Data is available for this title.
ISBN 978-1933771-80-9

Proofreading by Erica Lovett
Cover design by Laura Watkins
Text design and composition by PerfecType, Nashville, TN
Printed by Bang Printing

Distributed by Perseus Distribution
perseusdistribution.com

To place orders through Perseus Distribution:
Tel: (800) 343-4499
Fax: (800) 351-5073
E-mail: orderentry@perseusbooks.com

Significant discounts for bulk sales are available. Please contact Glenn Yeffeth at glenn@benbellabooks.com or (214) 750-3628.

TABLE OF CONTENTS

CHAPTER 1

Summer in Prague

The objects which astronomy discloses afford subjects of sublime contemplation, and tend to elevate the soul above vicious passions and groveling pursuits.

—Thomas Dick, introduction to
The Geography of the Heavens by Elijah Burritt, 1833

GURU: Today I will discourse upon the Violence in Astronomy.
DISCIPLE: Revered Sir! Will you be describing the violent phenomena in the Universe?
GURU: Yes, and I will also dwell upon the controversies amongst the astronomers about what these events imply—controversies which are no less violent than the phenomena themselves.

—Anonymous, quoted by Jayant Narlikar in
Violent Phenomena in the Universe, 1982

It was a hot summer day in Prague, but inside an air-conditioned auditorium in the Czech Republic capitol, tempers were even hotter. After loud and impassioned arguments, a group of astronomers had voted to reclassify Pluto from "planet" to a lesser status, and the decision did not sit well with the losing side. Henceforth, if the International Astronomical Union (IAU) were to have its way, Pluto would be a mere "dwarf planet," a new category established that day, August 24, 2006.

The IAU is a 90-year-old organization that promotes international cooperation in astronomy. As of 2009, 64 nations belong to it as "national members," and over 9,500 astronomers from 87 countries are individual members. The organization makes some decisions by vote of the national members, but some matters are put to a vote of the individual members who attend the IAU's General Assembly, a 10-day event that occurs, like soccer's World Cup, just once every 3 years.

Not every member who attends the General Assembly is there for the duration; the astronomers tend to come and go during the 10 days. The demotion of Pluto came on a vote on the last day of the General Assembly in 2006, which was attended by about 2,400 people, and by August 24, many of them had left Prague. Only 424 actually voted.

The adoption of the resolution that demoted Pluto was trumpeted around the world in what was surely the greatest media event in the history of the IAU. Other resolutions that were adopted on August 24, dealing with such topics as "precession theory" and "Barycentric Dynamical Time," were ignored by the press. Later in the day, the Assembly closed with a performance of medieval music and farewell remarks by the IAU's outgoing president.

It was made clear in the IAU decision that "dwarf planet" does *not* mean a particular type of planet. A dwarf person is still a person, a dwarf fruit tree—like a dwarf apple—is clearly a tree, but a dwarf planet is considered to belong to a different species than "planet," like "comet" or "asteroid."

Within weeks after its demotion, Pluto was given the numerical designation 134340, much as a convicted criminal is assigned an inmate identification number when he is incarcerated. The catalog number was issued by the Minor Planet Center (MPC), an arm of the IAU in Cambridge, Massachusetts. Most of the more than 134,000 objects in the Minor Planet Center's catalog of small bodies with well-determined orbital properties are, well, minor planets (the technical term for asteroids), but dwarf planets like Pluto get lumped in there, too. The MPC issues a monthly circular concerning the small bodies, usually near the time of full moon.

The edict on Pluto also displeased many other people who were not present and who, in many cases, were not astronomers or scientists of any kind. Unusually, for what seemed to be a mere matter of definition, the reclassification of Pluto from planet to dwarf planet outraged a great many people and amused others. And it stimulated numerous protests—some of them serious, some satirical, and some seemingly meant to heap ridicule on the IAU, a respectable scholarly organization that was previously unknown to most of the public.

In the United States, the general sentiment of the public and the news media overwhelmingly favored retaining Pluto as a planet. Pluto is popular with American schoolchildren, and supporting its planetary status does not involve new taxes or infringe on the constitutional right to keep and bear arms or any other socially divisive issue, so American politicians swung into action on Pluto's side in a rare display of bipartisan cooperation.

Two members of the state assembly in California introduced a bill, HR 36, denouncing the IAU decision, and about 50 of their fellow assemblymen and -women, listed alphabetically by last names from Aghazarian to Yee, signed on as coauthors. HR 36 pointed out in part that "[t]he mean-spirited International Astronomical Union decided on August 24, 2006, to disrespect Pluto by stripping Pluto of its planetary status and reclassifying it as a lowly dwarf planet." Referring to Mickey Mouse's pet dog, the bill noted that the astronomical "Pluto, named after the Roman God of the underworld and affectionately sharing the name of California's most famous animated dog, has a special connection to California history and culture" and complained, "The deletion of Pluto as a planet renders millions of text books, museum displays, and children's refrigerator art projects obsolete, and represents a substantial unfunded mandate that must be paid by dwindling Proposition 98 education funds, thereby harming California's children and widening its budget deficits."

In a similar, if more lighthearted, move, municipal leaders in the capital of Wisconsin introduced City of Madison Legislative File Number 04419, "Proclaiming Pluto as Madison's ninth planet." The tongue-in-cheek document criticized the IAU for failing to respect a diversity among planets in both size and gender preference, and invited "representatives of Pluto (should any exist) to an upcoming Common Council meeting where they will receive the keys to the city, passes to the Goodman Pool, an honorary membership in the Chamber of Commerce (to ensure that the folks on Pluto realize how business-friendly Madison really, REALLY is) and, finally, an autographed CD from the Dixie Chicks."

State Representative Joni Marie Gutierrez (Democrat), a landscape architect from Dona Ana County, New Mexico, introduced a bill that was adopted as House Joint Memorial 54 during the first legislative session of 2007, which concluded, "NOW, THEREFORE, BE IT RESOLVED BY THE LEGISLATURE OF THE STATE OF NEW MEXICO that, as Pluto passes overhead through New Mexico's excellent night skies, it be declared a planet and that March 13, 2007 be declared 'Pluto Planet Day' at the legislature."

Pluto enjoys especially great sympathy in New Mexico, because the discoverer of the little planet, Clyde Tombaugh, spent much of his life there, both at work and in retirement. So it was hardly surprising that astronomers, students, and relatives of Tombaugh (who had died in 1997, already distressed at prior attempts to demote Pluto) rallied at New Mexico State University in Las Cruces, where Tombaugh's papers are archived. They brandished placards marked "Protest for Pluto" and "Size Doesn't Matter." In Animas, New Mexico, about a two-hour drive from Las Cruces, a new real estate development for amateur astronomers and others who want to reside under dark skies suitable for stargazing, installed a covenant, according to Tombaugh's biographer and astronomer friend, David H. Levy, "that people living at Rancho Hidalgo must accept Pluto as a major planet."

The growing protests were not all rooted in sentiment or in making fun of stuffy scientists. Some were serious reactions to the IAU decision by experts who found fault with it.

Three hundred scientists, including prominent authorities on the solar system, endorsed a brief and pointed petition stating in full: "We, as planetary scientists and astronomers, do not agree with the IAU's definition of a planet, nor will we use it. A better definition is needed." The group began plans for a conference that would seek agreement on a much better definition of "planet" that might (or might not) restore Pluto to full planet-hood.

When the editor of a forthcoming professional encyclopedia of the solar system instructed authors to revise their chapters to follow the IAU's definition of what is and is not a planet, some scientists withdrew their manuscripts rather than comply.

Dr. Alan Stern, the chief scientist of NASA's New Horizons mission, a space probe en route to Pluto, took his cause to the people, announcing on the Internet that "[t]he New Horizons project, like a growing number of the public, and many hundreds if not thousands of professional research astronomers and planetary scientists, will not recognize the IAU's planet definition resolution of Aug. 24, 2006. . . . So on this Web site and in documents, discussions and other aspects of the New Horizons mission, we will continue to refer to Pluto as the ninth planet. I think most of you will agree with that decision and cheer us on." Stern has also authored a book on Pluto, done much research on Pluto's atmosphere and surface properties, and helped discover two of its three known moons. His probe will reach Pluto—regardless of how you define it—in July 2015.

Two professors of information systems at universities in Newfoundland and British Columbia, Jeffrey Parsons and Yair Wand, published an essay in the scientific periodical *Nature*, asserting that "the controversy over a sole correct definition of 'planet,' and whether Pluto falls within it, is unwarranted from a scientific perspective." They cautioned that "scientists should recognize that classification happens in the mind and, as a result, it can be influenced by beliefs and emotions." Pluto is what it is, regardless of how we pigeonhole it.

Mark Bullock, the director of the Center for Space Exploration Policy Research, a joint undertaking of a space science group in Colorado and a university department of philosophy and religion studies in Texas, asked "who has the social mandate to alter the definition of something as fundamental as a planet." He noted that "the word 'planet' has cultural, historical, and social meaning and as such requires much broader discussion and consensus than those required for the naming of astronomical bodies." In other words, he questioned the very mandate for the IAU to redefine "planet" and thereby demote Pluto. He may have had a good point. After all, in 1818 a New York City jury ruled that a whale is a fish. Their decision had the force of law, compelling a vendor of whale oil to submit his product to state inspection (and pay for the inspection), as required by New York State law for fish oil. The fact that biologists knew perfectly well that whales are mammals, not fish, had no legal effect. Sometimes official decisions are just plain wrong or at least a little bit fishy.

David A. Weintraub, a Vanderbilt University professor who wrote the book *Is Pluto a Planet?*, complained in the *Chronicle of Higher Education* that, in reaching their decision at the IAU General Assembly, "the process that astronomers followed was embarrassing, unkempt, and overly politicized."

One learned society reacted to the IAU decision in a manner that attracted much attention, without actually taking a position on the status of Pluto. The American Dialect Society (ADS), which has been a watchword in its field since 1889, voted to select "plutoed" as their 2006 Word of the Year. They announced that "[t]o pluto is to demote or devalue someone or something, as happened to the former planet Pluto when the General Assembly of the International Astronomical Union decided Pluto no longer met its definition of a planet." In the ADS tally, plutoed won a close victory (57 to 43) over the term "climate canary," after other contenders were eliminated. (A climate canary is "an organism or species whose poor health or declining numbers hint at a larger environmental catastrophe on the horizon.")

Then there was the larger public reaction, beyond academia and government. In some areas, Pluto supporters cruised the roadways with bumper stickers asking other motorists to "Honk if Pluto is still a planet." Newspapers had a field day, running articles and commentary under headlines such as

ADIEU, POOR PLUTO, SENT TO THE DOGHOUSE
(*The Age*, Melbourne, Australia)

WITH PLUTO GONE, WHICH OF US WILL BE NEXT?
(*The Toronto Star*)

GOING 'ROUND AND 'ROUND ON DEFINING PLUTO
(*The Boston Globe*)

ASTRONOMERS GOOFED ON PLUTO
(*The Times Union*, Albany, New York)

On the Internet, blogs were alive with controversy. Those opposing the Pluto demotion seemed amused by what some deemed incompetence and elitism among scientists. According to one blog active at the time (freemantlebiz.livejournal.com), "these parasitical life forms have been meeting in Prague, mostly at their respective countries' taxpayers' expense to spend, amongst other matters, untold hours thrashing out the semantics of whether or not Pluto should be ranked as the ninth planet. . . . [A]t the end of the day, all the bigwig astronomers are concerned about is their tenure, which translates as fat salaries, professional status, and travel perks." Other commentators supported the decision. "Pluto's characterization has bugged me since I was an early teen," wrote someone with the screen name Proesterchen in posted remarks on BadAstronomy.com, the leading American astronomy blog. Proesterchen added, "Today I can finally rejoice at its demotion. I drink to you, Pluto, the planet that never was." The Wikipedia page on Pluto was locked against additional contributions because of the contentious entries posted by pro-Pluto and anti-Pluto partisans. It has since been updated to reflect both the IAU decision on Pluto and the continuing opposition from certain astronomers and planetary scientists.

The Canadian indie rockers SubPlot A, self-described as "not so much a traditional band as a group of nomadic solo artists, musicians, wordsmiths, and soothsayers," recorded a song of protest, "Pluto Rocks," wailing, in part,

> "Pluto rocks, in a vacuous void
> Grandfather Pluto, he's no asteroid"

Jeff Mondak, a children's poet and songwriter in Champaign, Illinois, was inspired to compose his take on the controversy, "Pluto's Not a Planet Anymore." The song concludes with the catchy, if regretful, stanza

> "They met in Prague and voted
> Now Pluto's been demoted
> Oh, Pluto's not a planet anymore"

Teachers and textbook publishers wondered what to tell their students in the future about the population of the solar system. At Silverado Middle School in California, 21 of 32 seventh graders voted to keep Pluto's status as a major planet. Elsewhere, children wrote letters of protest to Neil deGrasse Tyson, the director of the Hayden Planetarium in New York City and a long-time proponent of demoting Pluto. Emerson York, a third grader in Mansfield, Pennsylvania, hectored Tyson, "Why do you think Pluto is no longer a planet? I do not like your anser [sic]!!!" The letter is reproduced with others in Tyson's *The Pluto Files: The Rise and Fall of America's Favorite Planet* (W.W. Norton and Co., 2009). Educators, who used to employ the mnemonic "My Very Educated Mother Just Served Us Nine Pizzas" to help pupils remember the order of the planets in distance from the Sun (Mercury, Venus, Earth, Mars, Jupiter, Saturn, Uranus, Neptune, and Pluto), went searching for a new mnemonic that omits Pluto. Among the suggestions:

> My Very Excellent Mother Just Sent Us Noodles
> My Very Educated Mother Just Said, "Um, No"
> My Very Elegant Mnemonic Just Stops Under Nine

Protest mnemonics were also proposed:

My! Very Educated Morons Just Screwed Up
Numerous Planetariums
Many Very Earnest Men Just Snubbed Unfortunate
Ninth Planet

In Maine, initial defiance was later tempered with partial acquies-
cence with the IAU decision, all to the benefit of science education.
Kevin McCartney, the public voice of the Maine Solar System Model,
a 40-mile-long series of monuments built to display the 9 planets,
arranged in proportion to their respective average distances from the
Sun (which is also represented), announced that Pluto would not be
removed from the model. The Maine Solar System Model, billed as
being "built by the people of Aroostook County," runs along Route 1
from Presque Isle to Houlton, with scale models of the planets. The
"Earth" (a 5.5-inch-wide ball of painted fiberglass over Styrofoam)
is on the roadside grounds of a Chrysler, Jeep, and Dodge dealer,
Percy's Auto Sales, just 1 mile from the "Sun." "Pluto," a 1-inch white
ball, is fixed (along with a ball representing Pluto's largest moon,
Charon) to a wall at the Houlton Information Center. Eventually,
however, a *second* "Pluto"—this one described as a dwarf planet—
was added at a different point on the route. Planet "Pluto" remains in
Houlton, at a location corresponding to Pluto's average distance from
the Sun, while dwarf planet "Pluto" is at the Southern Aroostook
Agricultural Museum at a place roughly corresponding to Pluto's distance at the time that the New Horizons probe will visit it in 2015.
Two other dwarf planets were added, one far beyond planet "Pluto,"
thereby extending the range of the Maine model to Topsfield, about
95 miles down Route 1 from the "Sun." Public interest rose to the
point that there were Planet Head Days at the University of Maine at
Presque Isle (where McCartney is a geology professor), when volun-
teers' heads were shaved and painted to look like planets to support
breast cancer awareness and the New Horizons mission.

The Lakeview Museum of Arts and Sciences, in Peoria, Illinois, which bills itself as "home of the largest complete model of

the solar system in the world," accepted the IAU decision. It labeled its Pluto, stationed at Good's Furniture in Kewanee, 40 miles from the museum, as a "dwarf planet." The commercial world also was quick to adjust to the IAU dictate. In September 2007, blogger Jason Kottke reported that a recently purchased child's ball, decorated with images of the planets, omitted Pluto. He remedied this by adding a fat black dot annotated "Pluto" with a Sharpie marking pen.

AN AMERICAN PLANET?

Some commentators insist that the public interest in Pluto is hardly a global phenomenon, a point of view supported by the many examples of public protest described above, which in most cases originated in the United States. It may be that the sentiment for retaining Pluto as a full-fledged or major planet is indeed largely American. Neil deGrasse Tyson, in *The Pluto Files* and in multiple television and radio appearances, ascribed this American preference for Pluto to the coincidence between the planet's name and the name of the Disney animated cartoon character Pluto. Disney films have been shown around the world, but perhaps they resonate the most with audiences here at home, where they are engrained in our common culture. Millions of Americans know Pluto the cartoon character from viewing Disney films or reading comic books. It may be no coincidence that the University of Central Florida, a school located near Disney World, was the place where an assistant professor of physics and planetary science was booed by 300 students when, he reported, "I told my class that Pluto wasn't a planet anymore."

For decades after the discovery of Pluto, the people who visited planetariums and science museums or who read up-to-date books on astronomy or who studied the solar system in elementary school, secondary school, or college were taught that Pluto is the outermost planet of the solar system. Many absorbed that knowledge and are reluctant to give it up based on a vote rather than on a specific new discovery about Pluto. And many people, especially children, may simply sympathize with Pluto as the smallest planet, a kind of anthropomorphized celestial underdog.

The strong support for planet Pluto in the United States understandably goes far beyond the younger generation. There are many adult Americans—astronomers and others—who take pride in Pluto as the only planet discovered in America and who don't want to see that distinction lost. Among the nine planets as we knew them before August 24, 2006, six have been known since antiquity—no one is credited with discovering Jupiter, for example; it was already visible to the naked eye when humans first looked up at the night sky. But three of the nine were discovered. Each find was a historic moment in science. The English can take pride in Uranus, discovered there in 1781, during the reign of George III. France and England share credit for finding Neptune, however much they may dispute each other's claim to having done it first. So naturally Americans are proud of Pluto. That's especially true in places like Flagstaff, Arizona, where the discovery of Pluto has been perhaps the finest moment in the history of the Lowell Observatory, and in New Mexico, where, as mentioned previously, Pluto's discoverer, Clyde Tombaugh, later settled, taught, and spent his old age, and where he is well remembered.

The nonscientists who speak up for Pluto may be a small fraction of the U.S. citizenry, but they probably constitute a much larger proportion of the scientifically attentive population, meaning those who follow the news of science as it appears in the daily press and through other channels with special interest. So naturally they will be highly represented among those members of the public who speak, write, or blog on the matter of Pluto, and they are more likely to know that Pluto was discovered in the United States.

Some people cannot help wondering if the IAU vote on Pluto was swayed by anti-American sentiment among some of the astronomers in Prague. After all, the General Assembly occurred during the second term of the George W. Bush administration when resentment abroad over U.S. foreign policy was especially high even among traditional American allies. Scientists are supposed to put chauvinistic feelings aside when deciding a scientific question, but they can be subject to the same prejudices as everyone else. In fact, as IAU officials prepared for the Prague meeting, they considered the possibility that a decision demoting Pluto might be perceived in the media as an expression of anti-Americanism, according to a report by the IAU Press Officer, Lars Lindberg Christensen.

THE ASTROLOGERS' DILEMMA

Even astrologers reacted to the demotion of Pluto from major to dwarf planet.

We the authors, like all modern astronomers that we know of, maintain that there is not the slightest shred of truth in astrology, the ancient belief that the heavenly bodies have direct influences on our lives on Earth. But millions of other people worldwide continue to read, listen to, or even heed the dictates of astrologers. We wish there was a daily science column in every newspaper that runs a daily horoscope, but that dream of "equal space" may never be realized.

Since shortly after Pluto was found and deemed a planet, astrologers began to incorporate it in some of their deliberations. (In the early 1930s, for example, a German astrologer asserted that "[w]hen Pluto is in the IVth house and inharmoniously aspected, he creates fanaticism, intolerance, phantasy, isolation.") But once the IAU demoted Pluto, what were the astrologers to do? Generally speaking, they want to take account of all major solar system bodies in casting their horoscopes, while not having to deal with minor objects such as the 134,000 or so asteroids and other small bodies in the catalog of the IAU Minor Planet Center. Should they keep Pluto in their calculations or not? People who trust horoscopes were likewise confused. Richard Harris, a science correspondent for National Public Radio, fielded a call from a male listener who, prior to the IAU vote, had commissioned a personal horoscope from a professional astrologer. The astrologer asserted that because of Pluto's position in the sky at the time that the gentleman was born, the listener would never marry. In the wake of Pluto's demotion to lesser status, the man wanted to know if his marital prospects had improved. (We think he should have spent less time with astrology and listening to the radio and gotten out of the house and socialized or maybe logged on to an Internet dating site.)

Of course, if astrologers would now strike Pluto from their horoscopes, it might be taken as an admission that their previous work, which included Pluto, was wrong. However, Rob Tillett, a veteran astrologer who resides on the east coast of Australia, reports on his Web site that the Pluto demotion by the IAU "is substantially irrelevant to astrology." He maintains that "our own research and that of other modern astrologers demonstrates that astrologers and their clients ignore it [Pluto] at their peril." Molly Hall, described as a "writer-stargazer, and your Guide to Astrology on About.com," seems to concur. She explains that although Pluto has been reclassified as a dwarf planet, "astrologers still reckon it to be a powerful force on a collective and individual level." So their predictions should be as good (or bad) as they were before.

COSMIC CONTROVERSIES

The 2006 arguments in Prague were just the most recent debates over the correct description of Pluto. In fact, disagreements over Pluto, which in some sense began (as we describe in Chapter 6) even before it was discovered and named, follow in a long, but unhallowed, tradition of astronomical disputes over planets, extending back for centuries. The ancient Greeks, in teachings that were accepted in Europe well into the 17th century, held that all planets (and the stars) revolved around the Earth. The Earth was not a planet to them, but the Sun and Moon, supposedly revolving around Earth, *were* planets. By the time Galileo began studying the heavens with his telescopes in 1609—soon concluding that the Sun, not the Earth, was the center—these earlier teachings had been adopted as dogma by the Roman Catholic Church. Those who held to the contrary that the Earth moves around the Sun (and was thus a planet) were not just disputing other scholars, they were subject to the Holy Inquisition. They could be arrested and punished, and their books were liable to be banned.

As new planets were found with larger telescopes and new mathematical methods after Galileo's time, there were disputes over their nature, their proper names, and even whether they existed at all.

(Some did; some did not.) A few astronomers claimed to see a planet nearer to the Sun than Mercury. Nothing as large as that claimed planet, Vulcan, exists inside Mercury's orbit, as we now know, but astronomers seek much smaller bodies, the vulcanoids, that might be there.

In the early 19th century, astronomers began discovering the first asteroids in the region between the orbits of Mars and Jupiter. At first, these, too, were considered planets, and for a time, there were 11 known "planets," until asteroids were relegated to a lesser category.

A famous dispute over Neptune—whether it was discovered by a Frenchman or an Englishman—continues to the present day, although the first person to knowingly see it was a German. Many astronomers (including the authors) now believe that it actually was first glimpsed by an Italian, Galileo.

In the 20th century, astronomers began reporting the discovery of planets around other stars, and many of these reports were strongly criticized. There were no accepted cases until the century was nearly at its end. And there have been strong disagreements about the discoveries of new bodies orbiting the Sun at roughly Pluto's distance or even far beyond. Oddly enough, until recently these disputes often were about the actual or claimed planets themselves rather than the criteria used to classify them as planets. However, when the IAU demoted Pluto in 2006, it took that action in the context of establishing a definition of the term "planet." The new definition was crafted to exclude Pluto. A decade earlier, when astronomers had begun discovering exoplanets—the planets of stars beyond the Sun—some or all of the initial discoveries (not counting "discoveries" that turned out to be false) were criticized by experts who asserted that the newfound objects were not planets but rather were "brown dwarfs," a then-hypothetical category of objects more massive than planets but lighter than stars. In one case (as we explain in Chapter 10), rival scientists each claimed to have discovered a small body orbiting the same star. One observer insisted it was an exoplanet and the other maintained that it was the first known brown dwarf. But all their observations and the explanations of what they found were wrong. The object does not exist.

Much of the current dispute over the classification of Pluto is in fact an argument about the proper definition of "planet." Even astronomers who accept or are reconciled to the demotion of Pluto are often prone to disagree with the IAU definition of planet, as we discuss in Chapter 2. There are now hundreds of generally accepted planets, the vast majority being the planets of other stars. Yet they don't fit the IAU definition, which—amazingly to some experts and laypeople alike—explicitly excludes exoplanets. The IAU has defined what is a planet in our solar system but is silent on what constitutes a planet in the vast remainder of the universe.

Why have so many errors occurred in the annals of planet finding? In some cases, astronomers were working at the limit of available technology, making it hard to tell if a dimly sensed object was real or just what modern scientists call "noise"—a false signal or a case in which the object may be real but not recognized for what it really is. And in other cases, astronomers predicting the locations of new planets were employing untried mathematical methods not yet accepted by other scientists, or they were making mathematical predictions from faulty input data (such as imprecise determinations of the orbit of a known planet). False alarms might occur, but the same astronomer who erred in one such case was likely to do so again.

In this book, we examine the Pluto controversy and place it in context within the long history of scientific disputes on planets. We are both members of the IAU, and one of us (Marschall) attended the General Assembly in Prague and voted for the resolution that demoted Pluto. The other (Maran) stayed home in the United States and so could not cast a vote, although he opposed the decision. We've agreed to disagree on Pluto and to do our best to present the reader with a fair and balanced account of the Pluto controversy, which we regard as just one case among many cases of what David Finley of the National Radio Astronomy Observatory in Socorro, New Mexico, has referred to as "astronomers behaving badly." These may be exceptions to the general rule of cooperation in the cause of science, but they are interesting exceptions.

CHAPTER 2

The Great Pluto Debate and the Great Pluto Debaters

. . . he that filches from me my good name
Robs me of that which not enriches him,
And makes me poor indeed.

—William Shakespeare, *Othello*, 1603

NAMING NAMES

When the International Astronomical Union adopted an official definition of "planet" and explicitly excluded Pluto from that category, it was just the latest step of a long campaign to demote the little planet. This series of events even included an earlier official notice that the IAU was *not* going to reclassify Pluto.

Lost in the furor after the IAU decision was the reason *why* the IAU was making a new definition of "planet." After all, it didn't have a previous definition of the term or a definition for most categories of astronomical objects. There's no IAU definition of "star," for example, or "nebula" or "galaxy." Why "planet"? Because an administrative necessity arose requiring an official definition.

The IAU functions as an international consensus maker in astronomy. The consensus building began as early as 1922 at its first General Assembly, where the IAU adopted an official list of constellation names and their abbreviations. The list was needed because astronomers were discovering hundreds of new variable stars (stars that change perceptibly in brightness), and the official system for cataloging each new variable star is to give it a name that includes the constellation in which it's located. For example, V878 Cassiopeiae, known as V878 Cas for short, is a variable star in the constellation Cassiopeia. The IAU adopted "Cas" as the abbreviation for "Cassiopeia," and it adopted Cassiopeia as one of 88 official constellations. When you combine a star identification such as V878 with a constellation name, you use the Latin genitive case of that name, so Cassiopeia becomes Cassiopeiae.

Astronomers didn't always agree on the precise boundaries between constellations. So how could they name a variable star that was close to a disputed border? To solve this problem, the IAU eventually adopted official boundaries between the constellations. Astronomy artists are free to draw the constellation Ursa Major (the Great Bear) any way they like. But the border around the Great Bear is officially fixed.

The IAU provides standard names, abbreviations, and boundaries for constellations so that astronomers in all countries can refer to celestial objects and their positions in the sky in a common reference system. By the same token, the IAU has the internationally agreed responsibility to name topographic features on bodies in the solar system, such as the mountains and craters on the Moon and Mars. And it has a similar responsibility to name new objects in the solar system, like comets, asteroids, and moons, as they are discovered and confirmed.

When the IAU assigns a name to a planetary feature, it follows existing traditions or naming conventions. For example, the convention for craters on Mercury is to name them for famous historical figures in the arts; a crater on that planet was named Rembrandt in February 2009. Features of all kinds on Venus are named after women, both real and mythological (with one exception: Maxwell, the name of Venus's highest mountain, honors physicist James Clerk Maxwell and was grandfathered in). There is Mead Crater, for example, named for anthropologist Margaret Mead, and Aphrodite Terra (after the Greek goddess of love), which is the largest highland area on Venus. The conventions for Mars include naming large craters after scientists who studied the planet and naming small craters (less than about 40 miles wide) after towns (typically with populations of less than 100,000) on Earth.

There is an IAU naming convention for each kind of newly found solar system object. Comets, for example, are named after the people who discovered them, like Comet Hale-Bopp, discovered in 1995 by Alan Hale of New Mexico and Thomas Bopp of Arizona. David Levy, the famous comet hunter who is also Clyde Tombaugh's biographer, has his name on more than 20 comets and is looking for more. Asteroids (minor planets) are named for individuals who did *not* discover them, usually at the suggestion of the discoverer. Minor planet 3001 is named for Michelangelo, and 6701 honors Andy Warhol; 3623 bears the last name of Charlie Chaplin, and 26858 is named Misterrogers. There are asteroids named for musicians from Mozart to Frank Zappa to Elvis Presley (minor planet 17059 Elvis orbits the Sun once every three years and nine months). And minor planet 9768 is named Stephenmaran after one of the authors, who is fortunate to be the friend of a successful asteroid-and-comet hunter.

Planets are named after gods and traditionally are not assigned official IAU numbers. The Earth has been called the "third rock from the Sun" in popular American parlance but not "planet 3." Small, icy bodies beyond Neptune and Pluto, discovered since 1992, are "Kuiper Belt objects" (KBOs), as we explain in Chapter 9, and the IAU numbers them and names them after mythological figures. So long as the KBOs were much smaller than Pluto, no one confused them with planets. But as KBOs approaching Pluto in size were discovered, and then one even a little larger than Pluto was found, the

IAU had a dilemma. If something is comparable to Pluto out in Pluto's region of the solar system, should it be named like a planet or numbered and named like a KBO? How do you tell the difference? That's why the IAU needed an official definition for "planet."

For centuries "planet," from the ancient Greek for "wanderer," meant a luminous celestial object that moved across the starry background of the constellations. That meant Mercury, Venus, Mars, Jupiter, and Saturn but not Earth. The Earth was not considered a planet; it was thought to be stationary at the center of the universe. Planets, it was thought, moved around the Earth. And since the Sun and Moon track across the sky background, they were considered planets as well. So there were seven known planets in antiquity. Comets appeared from time to time, often moving rapidly across the sky, but they didn't count as planets, because back then they were considered optical phenomena in the Earth's atmosphere, like halos around the Moon at night or rainbows in the daytime.

By the 18th century, scientists understood that the Sun is at the center of the solar system, that the Earth is a planet of the Sun, and that the Moon is a satellite of the Earth. So the number of known planets was down to six—Mercury, Venus, Earth, Mars, Jupiter, and Saturn—at least until 1781. In that year, William Herschel discovered the planet we now call Uranus with a telescope. There was no IAU yet and no naming convention for new planets, because no new planet had ever been discovered before. Herschel lived in England, under the rule of George III, so he attempted to curry royal favor by naming the new planet after King George III, as we discuss at greater length in the next chapter. But astronomers outside the king's realm did not accept the name for the new planet and eventually a nonpolitical name from Greek mythology, Uranus, was adopted through common usage. There was no vote or official decision on accepting the name Uranus; it just happened.

By 1848, an American textbook, *A Compendium of Astronomy* by the Yale College professor Denison Olmsted, taught that there were 11 planets. This tally included the seven planets up to Uranus, plus the first four asteroids that were discovered starting in 1801. The definition of a planet was still simply a wanderer across the heavens, and Olmsted called the 11 objects "primary planets" to distinguish them from moons, which he referred to as "secondary planets." (Neptune had been discovered in 1846, but apparently Olmsted had not gotten word of it before he finished his book.)

The number of planets had grown to about ninety according to an anonymous "Connecticut pastor," who published his *Parish Astronomy* in 1867. The big increase was in the number of known asteroids, which were still regarded as planets. The New England pastor also pointed to the possibility of many other planets of a new kind—what we now call exoplanets—that might orbit around many of the stars seen in the night sky, a speculation that was remarkably ahead of its time; the first exoplanets were not discovered until the 1990s, over a century later. During the mid- and late 19th century, a planet was, by common usage, an object that moved across the sky, was not a comet, and was known to orbit the Sun. By the end of that century, astronomers tended to distinguish the eight known planets (Mercury to Neptune) from asteroids. There was no official definition of "planet"; nor was there much concern over this deficiency. "Planet" seemed to be like "pornography," which has been described as very hard to define, but you know it when you see it.

After Pluto's discovery in 1930, there was some concern initially as to whether it was a planet, but no one stopped to consider how to define that term. In 1962, when the astronomers Otto Struve and Velta Zebergs published their historical review, *Astronomy of the 20th Century*, they simply referred to Pluto as the "[n]inth planet of the solar system." There's no definition of "planet" in the book, although other simple terms like "orbit" and "Milky Way" are defined.

Forty years later, *21st Century Astronomy*, a leading college textbook by Jeff Hester of Arizona State University and seven coauthors, was published. It does give an explicit definition of "planet" as a "[l]arge body that orbits the Sun or other star [and] that shines only by light reflected from the Sun or star." By its publication date of 2002, however, Pluto's status was under discussion. A section of the text asks whether Pluto is a "Tiny Planet or Gigantic Comet?" The authors state that "[i]f discovered today, they [Pluto and its largest moon, Charon] would certainly be considered especially large KBOs rather than a planet and its moon."

CAMPAIGNING AGAINST PLUTO

The main argument of those who support the demotion of Pluto today—that it is just one of the largest of many small, icy bodies in the outer solar system and not a distinctive major planet in its own right—was foreshadowed 80 years ago in the same year that Pluto was discovered. In 1930, Frederick C. Leonard, an astronomer at the University of California, Los Angeles, asked, "Is it not likely that in Pluto there has come to light the *first* of a *series* of ultra-Neptunian bodies, the remaining members of which still await discovery but which are destined eventually to be detected?" By "ultra-Neptunian," he meant "beyond Neptune."

Leonard must have realized, or strongly suspected, that the prediction of Pluto and its location in the sky, which led Clyde Tombaugh to search for and find it, was flawed. This would mean that the discovery was accidental, so that if a distant object like Pluto happened to be where Tombaugh looked, there might be others located where he didn't look (and perhaps a bit dimmer than Pluto, so Tombaugh's observations might be incapable of detecting them anyway). Sometimes a scientific theory seems to successfully pass an experimental check, but that doesn't prove the theory; independent tests by other scientists using other methods may be needed. One laboratory's seeming miracle drug may be another lab's useless therapy. A brilliant observational discovery might just happen by good luck and not much more.

In fact, Leonard correctly surmised the existence of what is now called the Kuiper Belt from the evidence of a single object, the newly discovered Pluto. However, his simply written article, published in 1930 as *Astronomical Society of the Pacific Leaflet* Number 30, seems to have been ignored by other astronomers of the day. The first Kuiper Belt object to be discovered and recognized as such (1992 QB$_1$) was not reported until 1992. David Jewitt, an authority on comets and the outer solar system who codiscovered the KBO with Jane Luu, later wrote that before 1992, the only "observed member [of the Kuiper Belt] was Pluto, misleadingly given planetary status for a host of mostly socio-scientific reasons." Jewitt maintains that rather than advancing science, the interpretation of the newfound Pluto as a previously predicted planet hampered science. He considers that if Leonard's concept of Pluto as one of a population of many icy bodies beyond Neptune had been accepted when it was proposed, "our study of the structure of the solar system could have been advanced by many decades. . . ."

In a way, the denigration of Pluto began even before Clyde Tombaugh discovered it in 1930. In 1929, Tombaugh, still a young observatory assistant in Arizona, not long removed from his Kansas farm, was cautioned by a prominent astronomer about his search for a planet, "Young man, you are wasting your time." Tombaugh's distinguished visitor was sure that all the planets had been found. He was like the director of the U.S. Patent Office who supposedly proposed in 1899 that the office be closed since everything capable of being invented had been invented by then. The Patent Office story is probably apocryphal, but if you agree with the 2006 IAU decision on Pluto, and you don't count exoplanets, Tombaugh's visitor was right, although possibly for the wrong reason. Tombaugh didn't take the advice; he kept looking, and he found Pluto. Depending on how you look at it, Tombaugh found a planet, or he discovered the Kuiper Belt before it was predicted, or (if you think, as some astronomers do, that Pluto is a planet that was the first known member of the Kuiper Belt) he did both.

At any rate, after 1930, Pluto was widely accepted as a planet and described as such in school books, college texts, and planetarium shows. Fifty years went by before the first modern assaults on Pluto's status occurred in 1980. It was a siege that began early that year in Las Cruces, New Mexico, and that lasted more than ten times longer than the World War II siege of Leningrad. However, at Leningrad, the German forces who conducted the siege ultimately lost. In Pluto's case, the assault forces finally prevailed at the IAU General Assembly in 2006. Pluto's backers are continuing the fight, but they lost a landmark battle.

The leading participants in the debates over Pluto since 1980 are an eclectic assortment of astronomers, some of them adept at getting their message out through the media, some of them used to working through the committees and old-boy networks of the academic world, and some of them effective at both. Although there are many prominent planetary scientists who are women, these vocal debaters and verbal battlers are mostly men.

Brian Marsden is perhaps the prime instigator of the modern campaign to reclassify Pluto. A genial, bespectacled expert on the orbits and identities of solar system objects, he directed the IAU's Minor Planet Center for many years, where (among other duties) he assigned the official minor planet numbers to newly recognized objects. He seems to have begun the current controversy on Pluto with remarks made at a conference in Las Cruces, New Mexico, in February 1980. And he's no stranger to other astronomical brouhahas. In March 1998, he issued an official IAU Circular seeking additional observations of a recently found asteroid (1997 XF_{11})[1] that, he calculated, might approach the Earth at a distance too close for comfort on October 26, 2028. New observations would allow him to improve knowledge of the asteroid's orbit, hopefully confirming that it would miss the Earth. In fact, additional observations provided that confirmation almost immediately, but nevertheless, there was a media field day over the celestial alarm. Other asteroid experts voiced harsh criticism of Marsden for not holding the announcement for a day or two and thereby avoiding embarrassment to astronomers. We thought some criticism was reasonable, but much was arguably too severe. Some of the critics of Marsden's 1998 announcement had also objected to his earlier proposal to demote Pluto. Was it planetary payback time?

If Marsden gave the original impetus in scientific circles to demote Pluto, it was Neil deGrasse Tyson who became the public face of the demotion, at least in the United States. Tyson, the handsome

---

[1] The name is a coded designation given to newly discovered asteroids. Once an asteroid's orbit is firmly established by further observations, its discoverer is allowed to propose a name.

and charismatic astrophysicist who hosts the Public Broadcasting System's *NOVA scienceNOW* television program, directs the Hayden Planetarium at the American Museum of Natural History in New York City. The undisputed successor to the late Carl Sagan as the most recognizable living U.S. scientist, Tyson is also, in the words of author Dava Sobel, "America's champion spokesman for space." When the new planetarium opened in February 2000 on the site of the old one, the deliberate omission of Pluto from an exhibit of the planets led to an intense controversy that catapulted Tyson to the media-proclaimed mantle of "the man who demoted Pluto." After the IAU downgraded Pluto in 2006, and especially after Tyson's book, *The Pluto Files*, was published in 2009, he again was introduced as the "man responsible for demoting Pluto" as he made the rounds of late-night and comedy talk shows and serious broadcast programs, including those of National Public Radio. Tyson has spent much effort trying to understand why public reaction to the demotion is far more intense in the United States than elsewhere, concluding that it results from the popular appeal of the Disney animated character Pluto, Mickey Mouse's pet dog. "I blame it on the dog because people in Europe don't behave this way," Tyson remarked on Comedy Central's *The Daily Show* in January 2009.

Andrea Milani, a mathematician and expert on asteroid orbits at the University of Pisa, emerged as a pivotal figure at the General Assembly in Prague. He was among the most vocal critics of a resolution that was introduced to officially define "planet" while preserving Pluto's planetary status. The members of the IAU's Planet Definition Committee, who formulated this definition, sat slumped in their chairs according to *Nature* magazine, as critics—with Milani speaking first—lined up at microphones to "denounce the definition in tones ranging from offended to furious." Experts such as Milani, who are called dynamicists because they study the orbital motions, rotational properties, and mutual gravitational interactions of celestial bodies, were angry that they had not been represented in the deliberations of the Planet Definition Committee. At one point in the discussion of the committee's recommendations, according to a reporter for *Science* magazine, Milani "literally screamed" in disdain at the alleged lack of dynamics in the work of the committee. A colleague of ours who witnessed the goings on sent an e-mail from Prague describing the situation as "astronomers behaving badly."

If Marsden, Tyson, and Milani were the most prominent advocates of reclassifying Pluto as a dwarf planet, Mark Sykes and Alan Stern were and are Pluto's biggest supporters, the leaders of the opposition to the little planet's demotion and among the most severe critics of the IAU's new definition of "planet." That new definition was written during the General Assembly that adopted it when it became clear that the proposal of the Planet Definition Committee was unlikely to be accepted.

Sykes, a hearty, bearded astronomer who directs the Planetary Science Institute in Tucson, brought unique skills to the debate over Pluto. Besides having professional status as a planetary scientist who had chaired the American Astronomical Society's Division of Planetary Science, he is an attorney who has practiced before the federal district court in Arizona. He can defend Pluto or criticize a definition or resolution with the best of them. And as an operatic bass-baritone who has performed in 59 productions, from *Boris Godonov* to *La Traviata*, since 1984, his resonant voice projects through the length and breadth of any hall. Sykes began defending Pluto long before the IAU met in Prague and has kept up constant criticism since 2006, including organizing both an online petition and a conference to find a better "planet" definition.

Alan Stern is a red-haired astronomer, manager, pilot, and flight instructor whose accomplishments stand out even in the community of the leading planetary scientists. He is one of the most productive researchers studying Pluto and its atmosphere, surface, basic nature, and moons. By campaigning effectively with legislative leaders and the general public in the United States, he succeeded in compelling an initially reluctant NASA to implement the New Horizons space probe. That spacecraft is now en route to Pluto and the Kuiper Belt under his scientific direction. He founded and directed an important research organization on space science in Boulder, Colorado, and briefly served (after the launch of the Pluto probe) as the head of NASA's Science Mission Directorate in Washington, DC. In 2007,

Time magazine named Stern as one of the "100 Most Influential People in the World." Stern may be the most outspoken critic of the "planet" definition that the IAU adopted, alternately ridiculing the adopted wording and writing thoughtful essays on how he thinks "planet" should be defined.

Owen Gingerich, who chaired the IAU Planet Definition Committee, is a white-haired emeritus professor of the history of science and of astronomy at Harvard University. He's not just an eminent scholar on early studies of the solar system. He descends from the Ivory Tower to serve as an expert witness in legal proceedings and is a sought-after consultant on rare books. Gingerich is also that rare individual who has both received Poland's Order of Merit and delivered an Advent sermon at the National Cathedral in Washington, DC. He is a relative rarity among leading scientists today, someone who professes strong interest in religious thought. His 2004 work *The Book Nobody Read*, about experiences in tracking down and studying the rare surviving copies of Copernicus's great work *On the Revolutions of the Heavenly Spheres*, was followed by another book in 2005, a collection of his lectures titled *God's Universe*. The latter book might not endow him with a halo, but as you might expect, Gingerich's advice is usually treated with due respect. However, at the General Assembly in Prague, this was far from the case.

Michael A'Hearn, a tall, bearded astronomer at the University of Maryland in College Park, holds the rank of Distinguished University Professor but is as likely to be seen in shorts and a Hawaiian shirt as in more formal wear when he speaks at a NASA press conference or scientific conference. He thought from the outset that Brian Marsden's proposal to number Pluto in the list of minor planets would be accepted by the planetary community if Pluto could keep its status as a major planet as well, so that it would be classified in both categories. Thus he struck a middle ground between demoting Pluto and no action on its status at all. A'Hearn, who was president of the IAU division on Planetary Systems Sciences at the time of the Brian Marsden initiative, told us recently that "Brian and I explicitly discussed the question of dual classification—he was in favor of it and brought it up with me." A'Hearn thought that Pluto might be viewed as a transitional object between the major planets and the "ice dwarfs," a term that scientists were beginning to apply to bodies in the Kuiper Belt. To advance this point of view at a conference, he told us, "I gave a talk that was mostly about *Archaeopteryx* and the more than 100 years of argument about the transition between a dinosaur and a bird."

The movement to demote Pluto and redefine "planet" had been brewing at least since Marsden's overture in 1980. At one point in February 1999, the IAU, which had not yet reached a decision even to formulate a definition of "planet," felt compelled to issue a press release signed by the general secretary, Johannes Andersen. This statement assured the press and anxious public,

> No proposal to change the status of Pluto as the
> ninth planet in the solar system has been made by
> any division, Commission or Working Group of the
> IAU responsible for solar system science. Accord-
> ingly, no such initiative has been considered by the
> Officers or Executive Committee, who set the policy
> of the IAU itself.

The fuss building up to the Prague General Assembly flared up again on February 19, 2000, when the spectacular new Hayden Planetarium opened to the public. In a review of the facility published in the May 2000 issue of *Sky and Telescope* magazine, Stephen Maran wrote, "[O]n one exhibit wall, I noticed the glaring absence of Pluto among the other planets of our own and other planetary systems." Others would soon notice the omission and foment strong public objections. The events were reminiscent of the cancellation of an aircraft exhibit at the Smithsonian Institution's National Air and Space Museum in the mid-1990s after veterans groups and congressional representatives objected to the planned wording of exhibit texts that referred to certain scholars' doubts of the military necessity of dropping the first atomic bomb on Japan. In that case, the protesters triumphed, a new exhibit without the offending words was mounted,

U.S. Senate hearings were held, and the museum director (himself an astronomer) resigned.

But Neil Tyson, the Hayden director, had no intention of backing off on his Pluto-less exhibit. He had held the view that Pluto was no planet for some time and had expressed it in a 1999 article titled "Pluto's Honor" in *Natural History* magazine. The pressure on him intensified in January 2001 when the *New York Times* ran a front-page story headlined "Pluto's Not a Planet? Only in New York." The stream of criticism from the public, Plutophile scientists, and schoolchildren, as mentioned in Chapter 1, began. Tyson wrote in *The Pluto Files* that on the day of the *Times* story, his voice mail filled to the limit as calls on Pluto kept coming in, and his e-mail in-box was likewise overflowing. He added, "[M]y life would never be the same." This was followed by a visit from Mark Sykes, who argued with Tyson over Pluto, leading to another *Times* article accompanied by a gag photo (in both senses of the term) of Sykes with his hands on Tyson's throat.

In 2004, the IAU appointed a committee of 19 experts in planetary science, including several of the astronomers profiled above, to make a scientific determination of what properties define a planet. After extensive deliberations and the exchange of many e-mailed opinions, they were able to reach no agreement whatsoever. As these discussions continued into 2005, more new Kuiper Belt objects were being discovered, some approaching Pluto in size and one, provisionally listed as 2003 UB_{313}, even larger. The matter had come to a head. The IAU needed to decide how to name UB_{313}, which meant it needed to know whether to follow the naming convention for major planets or another convention. A new Planet Definition Committee was appointed, led by Owen Gingerich.

The new committee was to deliberate in secret to avoid fanning the flames of public controversy over Pluto. Instead of just planetary experts, it included people with a wider perspective on the historical and social implications of the concept of "planet." Notably, one member was a non-astronomer, Dava Sobel. Sobel is an independent scholar who is the best-selling author of *Longitude* and *Galileo's Daughter*, works written for the general public that are well-founded in historical research. Even this smaller, more sedate group, which included the chair of the earlier 19-member committee and had the sobering presence of Catherine Cesarsky, the high-powered president-elect of the IAU, had problems reaching a consensus in time for presentation to the General Assembly in Prague, but it did do so.

Meeting in Paris in late June and early July 2006, the Planet Definition Committee, according to Gingerich, "had vigorous discussions of both the scientific and the cultural/historical issues." He recalled that some members lost sleep "worrying that we would not be able to reach a consensus." They were under pressure to submit a definition in time to be included on the IAU agenda for the August meeting. And they finally did reach a unanimous decision. Here is their definition: "A planet is a celestial body that (a) has sufficient mass for its self-gravity to overcome rigid body forces so that it assumes a hydrostatic equilibrium (nearly round) shape, and (b) is in orbit around a star, and is neither a star nor a satellite of another planet."

To avoid a public controversy over this proposed definition (under which Pluto would continue to qualify as a planet) in the few weeks remaining before the General Assembly met, it was kept secret until announced in an IAU press release on August 16, just after the meeting began. Besides Pluto, the largest asteroid, Ceres, would be granted planet status by this definition, as would Pluto's biggest moon, Charon, and the recently discovered UB_{313}. In other words, there would now be 12 known planets in the solar system, with the prospect of more to come as astronomers searched for other objects like UB_{313} in the Kuiper Belt. Pluto and Charon were considered a double planet in the deliberations leading to the proposed definition, because the center of mass of the two objects is outside the larger one. In contrast, for example, the center of mass of the Earth-Moon system is inside the Earth. Charon orbits around a point in empty space, while the Moon orbits a point that is within the Earth.

As mentioned earlier, the committee's proposed definition was violently criticized at the General Assembly. That's very unusual in the proceedings of this usually sedate meeting. Some astronomers—such as Andrea Milani—objected to exclusion of dynamical considerations from the criteria of the proposed definition. Perhaps they would rather have had one of their own experts on the committee, instead of a writer, a historian, or an IAU officer who was not a planetary specialist. Other astronomers in the Assembly were uncomfortable with the potential for many additions to the list of solar system planets as new large KBOs were discovered. And many were angry that the proposed definition had been kept secret until the meeting began. The Planet Definition Committee "so far as I know never solicited input," A'Hearn told us, adding, "The IAU membership heard about [the proposed definition] about the same time that the press did." It seems that there was something for everyone to dislike in either the proposed definition or the process by which it was formulated. In the end, the committee's definition was never put to a vote.

After heated discussions in the days following the opening of the General Assembly, a new definition was composed to meet the demands of the dynamicists, and on the last day of the meeting, when most participants of the Assembly (including Gingerich) had gone, the remaining attendees adopted the new "planet" definition, whose criteria now excluded Pluto, Ceres, Charon, and UB_{313}. The solar system was back to just eight planets, as it had been until 1930 when Tombaugh discovered Pluto.

Here's what the IAU adopted (in part):

> (1) A planet is a celestial body that (a) is in orbit around the Sun, (b) has sufficient mass for its self-gravity to overcome rigid body forces so that it assumes a hydrostatic equilibrium (nearly round) shape, and (c) has cleared the neighbourhood around its orbit.

> (2) A dwarf planet is a celestial body that (a) is in orbit around the Sun, (b) has sufficient mass for its self-gravity to overcome rigid body forces so that it assumes a hydrostatic equilibrium (nearly round)

shape, (c) has not cleared the neighbourhood around its orbit, and (d) is not a satellite.

(3) All other objects orbiting the Sun shall be referred to collectively as "Small Solar System Bodies."

There was no official record made of the vote, which the IAU later described as "a great majority." The vote was taken by a show of hands or, more accurately, a show of yellow hand paddles, like the bidding at an auction.

To leave no wiggle room for Plutophiles, this resolution as adopted included a footnote that explicitly stated: "The eight planets are: Mercury, Venus, Earth, Mars, Jupiter, Saturn, Uranus, and Neptune." And to further hammer home the point that Pluto was demoted, another decision was made by a vote of 237 to 157, with 17 astronomers abstaining: "The IAU further resolves: Pluto is a 'dwarf planet' by the above definition and is recognized as the prototype of a new category of trans-Neptunian objects."

The part of the adopted planet definition that has most confused the public and raised the most objections from astronomers who dislike the definition is part (c). That is the requirement that to be a planet, a body must have "cleared the neighbourhood around its orbit." Neighborhood is a vague term. Pluto certainly has not cleared its neighborhood of the Kuiper Belt, since thousands of KBOs remain there. But even a body as large as the Earth, if put in Pluto's location, would not be able to swallow up or gravitationally eject the great majority of the KBOs. And, as some experts point out, there are large numbers of so-called Trojan asteroids clustered at certain locations on Jupiter's orbit. But no one would say Jupiter is not a planet.

The demotion of Pluto led to enormous publicity and numerous complaints, as described in Chapter 1. They continued through February and March 2009 during the writing of this book. Most notably, on February 26, 2009, the Illinois State Senate, proud that Clyde Tombaugh was born on a farm near Streator, Illinois, adopted a resolution that stated that "Pluto was unfairly downgraded to a 'dwarf planet.'" It concluded:

RESOLVED, BY THE SENATE OF THE NINETY-SIXTH GENERAL ASSEMBLY OF THE STATE OF ILLINOIS, that as Pluto passes overhead through Illinois' night skies, that it be reestablished with full planetary status, and that March 13, 2009 be declared 'Pluto Day' in the State of Illinois in honor of the date its discovery was announced in 1930.

Many scientists continue to dislike the definition of planet that the IAU adopted in 2006, whether they consider Pluto a planet or not. This led to a conference on how to make a better definition, held in August 2008 at the Applied Physics Laboratory of the Johns Hopkins University in Laurel, Maryland, attended by many astronomers and science educators. There were erudite lectures, panel talks, much polite commentary from the audience, and a debate between Tyson and Sykes, moderated by a noted science broadcaster, Ira Flatow. Sykes kept his hands off Tyson's throat, but Tyson interrupted and spoke over Sykes with practiced ease. In the end, most attendees seemed to agree that the IAU had not defined "planet" satisfactorily. But they could not agree on a better definition, and so the prospect for the future is that the controversy over Pluto will continue.

To be fair, the scientists who are wrangling at present over the definition of the term "planet" and the status of Pluto are not just arguing over semantics. They are deeply concerned about the state of planetary science and want to make sure that scientific terminology accurately represents what we know about the structure, dynamics, and origin of the various bodies in the solar system. Still, it's hard not to feel a sense of déjà vu. Pluto was not the first planet to stir up a tempest among astronomers, as we shall see when we look back at past planetary discoveries. Nor, if history is a reliable guide, will it be the last.

CHAPTER 3

Contentious Planets: The Early History of Planetary Disputes

> I should disclose and publish to the world the occasion of discovering and observing four Planets, never seen from the very beginning of the world up to our own times. . . .
>
> —Galileo Galilei, *Sidereus Nuncius*, 1610

The discovery of a new member of the solar system invariably creates ripples of controversy, like a gold nugget brought into a saloon by a prospector. Is the find real, fraudulent, or illusory? What is its significance? Who gets to claim it, profit from it, or even name it? Those tables of planets, satellites, and asteroids you find in modern astronomy textbooks neatly encapsulate our current knowledge of the Sun's family, but they hardly do justice to the uncertainty and dispute that once swirled around astronomy's most important discoveries. The stories of Jupiter's moons and of the discovery of Uranus are two cases in point.

THE FIRST DISPUTED DISCOVERY

Prior to 1609, when Galileo first turned a telescope skyward, controversies over new planets simply did not exist, because no one ever discovered *any* new heavenly bodies. That's not to say that the heavens, as seen by the unaided eye, never had unexpected visitors. Far from it: meteors slashed the blackness every night; bright fireballs (especially bright meteors) appeared every now and then, even in daytime; and few people lived their lives without seeing at least one bright comet gracing the nighttime sky. More rarely, a new star, or *nova* (today understood as an exploding star), would flash briefly into view. Then, as now, there was always something new under the stars. But though people saw changes overhead, they did not recognize them as having any astronomical significance. Meteors, comets, fireballs, and novae were regarded, at least by educated people, as mere atmospheric phenomena.

According to the teachings of Aristotle, the ultimate reference source in the West for almost two millennia, the heavens were, by their very nature, eternal and changeless. The stars were fixed with respect to one another, as if painted on the sky, and though the constellations circled around the heavens once a day, they always returned to the same place. There were only seven planets, as we noted in the last chapter, that moved with respect to the stars—the Sun, the Moon, Mercury, Venus, Mars, Jupiter, and Saturn—and though their motions were more complex than those of the stars, they, too, seemed to repeat the same patterns over and over, if one waited long enough.

Aristotle envisioned the universe as a series of concentric crystalline spheres carrying the planets and stars around the stationary Earth, located at center of it all. The first sphere above the Earth carried the Moon; then came the inner planets Mercury and Venus; then the Sun; then the outer planets Mars, Jupiter, and Saturn; and finally an outermost sphere, which carried the stars. The planets themselves, which rode around on their crystal carriers like inlaid jewels, were perfectly smooth spheres, destined to remain that way forever. They were subject to no change at all because they were composed of a special substance called aether.

The only place in the universe where change took place, Aristotle taught, was in the region inside the crystal sphere that carried the Moon—the Earth and its atmosphere. In this sublunar, terrestrial realm, everything was made of four elements: earth, air, fire, and water. Rather than traveling in circles like the aetherial things in the heavens, these elements naturally moved in straight lines toward or away from the center of the universe—the center of the Earth. Water and earth, left to themselves, would sink down, and air and fire would move up. Anyone could see the truth of this for himself, for spilled wine—made of watery stuff—always dripped downwards, and the flame of a candle—made of fiery stuff—always pointed up.

So when someone saw an unfamiliar light in the sky, the flash of a meteor or the glow of a comet, it was clearly not to be regarded as something beyond the sphere of the Earth, but rather as a luminous substance floating in the Earth's atmosphere. Comets, for instance, are known today to be small balls of gravel and ice that move in highly elongated orbits and that eject long tails of gas when they come close to the Sun. But they were described by Aristotle as clouds of vapor released from the Earth by the heat of the Sun. As these clouds rose up through the atmosphere, they were set afire by friction and then moved around the sky by the rotation of the heavenly spheres above them. If you had gone to school in the time of Galileo, you would have learned to regard comets as terrestrial events, in accord with Aristotle's teachings. Your parents might have cautioned you, also, that comets foretold bad luck, because there was a persistent popular belief that comets, wherever they were located, were omens of doom.

So it was that the first discovery of previously unknown heavenly bodies had to wait until 1609. In that year, Galileo Galilei, a professor of mathematics in Pisa, caught word of a new invention, a tube with lenses at each end that made distant objects appear closer. He called it the *perspicillum*, or spyglass; we call it the telescope. Galileo was an expert craftsman, and he thought he might turn a tidy profit by designing an improved telescope and selling his device to military men and merchants, who could make good use of its ability to make invisible things visible. But in the fall of 1609, when he pointed one of his new telescopes to the heavens, he realized that that he held in his hands a tool that would forever change the old Aristotelian view of the universe. Overnight, Galileo became an astronomer.

Just a few months later, Galileo published a short book, *Sidereus Nuncius*, or the "Starry Messenger," describing some of his most striking astronomical discoveries. He had first looked at the Moon and was surprised to find that it was not the perfectly smooth ball that Aristotle described, but a world like the Earth, covered with rugged mountains and pockmarked here and there by craters. The telescope also revealed that the planet Venus showed phases like the Moon and that the light of the Milky Way came from vast numbers of individual stars, so closely packed together that they looked like a glowing river to the unaided eye.

By far the most remarkable discovery revealed by the telescope was that the planet Jupiter was accompanied by four moons, which circled around it like a quartet of leashed puppies, as it moved through the sky. Galileo had noticed the moons almost immediately after pointing his telescope at Jupiter; they looked like little stars, albeit curiously strung out in a line on opposite sides of the planet. It was only after watching for a few nights that Galileo realized what was going on: the four "stars" were actually smaller bodies in orbit around Jupiter. He continued to trace their motions and to measure how long it took them to go around the planet. The closest moon took a little under three days, and the most distant took a little over two weeks.

There was no word to describe these new worlds. Galileo, as you may note in the epigraph to this chapter, called them "planets." The four moons of Jupiter, which we now call the "Galilean Satellites," were, in fact, the first new objects since prehistoric times to be recognized as heavenly bodies and the first to orbit any body other than the Earth. Galileo, however, understood them as something far more significant—proof that the Earth was not the center of the universe, but moved around the Sun just like the planets. In 1543, a

Polish cleric named Nicolaus Copernicus had published a book, *De Revolutionibus Orbium Coelestium* (*On the Revolutions of the Heavenly Spheres*), in which he argued that the Sun was the center of the solar system, that the Earth was a planet, and that the only celestial body that went around the Earth was the Moon.

Galileo, like many European astronomers of the time, found Copernicus's Sun-centered universe appealing, but he had no way of answering some of the more pointed Aristotelian arguments against it. "How could the Earth go around the Sun," argued the keepers of conventional wisdom, "without leaving the Moon behind?" Jupiter's moons, which were clearly being carried around by Jupiter, provided the answer Galileo was looking for. Just look for yourself, he said: Jupiter's moons weren't left behind as it moved through the heavens, no matter whether it circled the Sun or the Earth. If Jupiter's moons could keep up with their planet, then surely the Moon could keep up with the Earth. "We have," he wrote in *Sidereus Nuncius*, "an excellent and splendid argument for taking away the scruples of those who . . . are so disturbed by the attendance of one Moon around the Earth while the two together complete the annual orb around the Sun that they conclude that this constitution of the Universe must be overthrown as impossible."

HOSTILE WITNESSES

Though Galileo wrote with clarity and conviction, there were still some skeptics who refused to accept the discovery at face value— quite understandably, if you see things from the perspective of an intellectual of the 17th century. To them, the telescope was a novel and largely unfamiliar device. People didn't immediately associate it with astronomy, as we do today. Prudence, not foolishness, led them to distrust what the telescope revealed, especially when those revelations were so at odds with accepted truths. Scholar Martin Horky expressed a typical reaction in a letter to the German astronomer Johannes Kepler shortly after the publication of *Sidereus Nuncius*: "It is a marvelous thing, a stupendous thing; whether it is true or false, I know not."

Even those who looked for themselves were not certain what it was that they were seeing. Galileo was bemused by a letter that described the experience of one of his friends, Cesare Cremonini, who complained, "I do not wish to approve of claims about which I do not have any knowledge, and about things which I have not seen . . . to observe through those glasses gives me a headache. Enough!" Galileo cautioned users that you couldn't just slap together two lenses and a tube and expect to see anything of significance in the heavens. It took carefully constructed optics, a steady mount, and dark skies to see what he had described in his book and to avoid getting a headache.

Granted, some staunch Aristotelians were just being obstinate. Galileo expressed astonishment, in a letter to Kepler, that there were those who refused to look through the telescope at all. "What shall we make of all this?" he groused. "Shall we laugh or shall we cry?" And he had little patience with Giulio Libri, a professor of Aristotelian philosophy who had been an irritant to Galileo since the two had been together on the faculty at the University of Pisa. When Libri died, not long after *Sidereus Nuncius* appeared, Galileo snidely remarked in a letter that the old man, not having wanted to see the moons of Jupiter from Earth, might finally get a chance to see them on the way to heaven.

Still, there were well-meaning intellectuals, especially those with a scientific bent, who realized that caution was in order. Was it not possible that spurious reflections inside the telescope could fool the observer into thinking that Jupiter was flanked by moons when, in fact, there were none? Christopher Clavius, a noted Jesuit astronomer in Rome, joked in a letter to Galileo that, to see the moons of Jupiter, one would have to first build a spyglass that creates them. But Clavius was not averse to trying things out for himself, and he and Galileo corresponded at some length on how to build and use a telescope effectively. Late in 1610, after some work, one of Clavius's Jesuit colleagues at the Collegio Romano, Giovanni Paolo Lembo, managed to build a quality telescope that did, in fact, permit the viewing of Jupiter's moons.

By the end of the year, the Roman astronomers were charting the motions of the moons themselves. In March 1611, Galileo visited the Jesuits in Rome and spent some time with Clavius. "I found that these fathers, having finally recognized the truth of the new Medicean planets, have been observing them here continuously for two months and continue to do so." The Jesuit observations, he noted, agreed well with his own sketches of the satellites. They may not yet have agreed on the interpretation—the Jesuits were not yet ready to concede that the Earth-centered universe must yield to the Sun-centered model of Copernicus—but they were in complete accord that the telescope worked as advertised and that the moons of Jupiter were not mere optical trickery.

THE NAME GAME

Jupiter's newfound moons raised a question of a more practical nature, a question that was to be raised in the future every time a new solar system object was discovered: what should these objects be named? Since no one had ever discovered a new member of the solar system before 1610, no procedure had yet been established for naming either moons or planets, and Galileo simply seized the prerogative for himself. On the title page of his book, *Sidereus Nuncius*, he announced the discovery of "FOUR PLANETS swiftly revolving around Jupiter at differing distances and periods, and known to no one before the Author recently perceived them and decided that they should be named the MEDICEAN STARS." Galileo's intention was to name them after his patrons, the Medici family, and this is why he called the stars "Medicean." He did this, in part, out of a sense of gratitude, for Ferdinand de' Medici, Grand Duke of Tuscany, had arranged for his appointment to the University of Pisa 20 years earlier.

A more important factor in Galileo's choice of name, it is certain, was his desire to better himself politically and financially. The Medicis, rulers of Tuscany, were the most powerful family in Italy both politically and financially, and though they had a history of ruthlessness toward those who threatened them, they were also known as great patrons of art and science. In 1609, the Grand Duke of Tuscany was Cosimo II, Ferdinand's son, whom Galileo had tutored in mathematics when he was younger. Honoring Cosimo de' Medici with four stars in the heavens, Galileo thought, might result in the granting of substantial favors.

The flattery worked: Cosimo remembered his old teacher with fondness and offered Galileo the post of chief mathematician and philosopher to the Medicis in Florence. The honorific came with a substantial stipend that allowed him to continue writing and research without the day-to-day worry of giving lectures at the university.

Suggesting that Jupiter's moons be named in honor of the Medicis may have been good for Galileo's pocketbook, but it had little effect on the actual names astronomers applied to the four new worlds. Simon Marius, a Dutch astronomer who had observed the moons around the same time as Galileo, claimed that he deserved priority for the discovery and that therefore he had the right to name them after his patron, the Duke of Brandenburg. Astronomers of other nationalities, who had vested interests neither in Florence nor Brandenburg, were not impressed and simply referred to the satellites by number, designating the closest one as Jupiter I and the most distant Jupiter IV.

Roman numerals worked well enough for a few hundred years, but in the 1890s and early 1900s, with better telescopes at their disposal, astronomers began to discover additional smaller moons of Jupiter, some closer to the planet than Jupiter I and some farther out than Jupiter IV. As the list of Jovian satellites grew, it became difficult to retain the old numbers and still maintain a sensible numerical sequence, so astronomers began casting about once again for names. Among the proposed naming schemes was one originally suggested by Simon Marius as an alternate to "Brandenburg" back in the 1600s. Half in jest, perhaps, he had mentioned to Johannes Kepler that the four moons, which circled Jupiter like ardent admirers, be named after four of Jupiter's lovers: three girls, Io, Europa, and Callisto, and one young boy, Ganymede. Names from classic mythology appealed to the Victorian sensibility, even given the sexual overtones of these, and the names stuck. Though there was no organization that could give an official stamp to the naming, people began to use the classic names in journal articles and textbooks, and as more moons of Jupiter were discovered later in the 20th century, they, too, were named after others in Jupiter's (quite large) entourage of paramours.

Galileo's failure to prevail in the naming of the moons he had discovered set a precedent that was followed until the 20th century, when a naming authority of record was finally established. Before the founding of the International Astronomical Union, discoverers of new solar system objects could advise on a name, but popular usage, in the end, would decide what name became "official."

URANUS: THE FIRST NEW PLANET

Not until 1781, with the discovery of the planet Uranus, did the telescope reveal anything to rival Galileo's discovery of Jupiter's moons. Why did it take so long? In many respects, Galileo's pioneering work, brief as it was, had almost pushed the telescope to the limit, pending major improvements on its ability to collect light and to resolve fine detail in the sky. In the 1600s and 1700s, telescopes grew in size, but they were unwieldy and difficult to use, and their optical design only allowed astronomers to look at small patches of sky at any one time. Not surprisingly, many things that might have been discovered went unnoticed. Discoveries required not only great skill in manipulating the giant optical devices, but also a large amount of luck.

By the mid 1700s, the only objects that had been added to the roster of the solar system were five moons of Saturn. The first was the giant moon that we know today as Titan, the second-largest satellite in the solar system (the largest is Ganymede). Titan was discovered by the Dutch astronomer Christiaan Huygens in 1655, who referred to it simply as *Saturni Luna* ("Saturn's Moon" in Latin). Four smaller moons were discovered two decades later by the French-Italian astronomer Jean-Dominique Cassini: Iapetus in 1671, Rhea in 1672, and Dione and Tethys in 1684. Like Galileo, Cassini wanted to name the four moons he had discovered after his patron, Louis XIV, but again, international acceptance was not readily forthcoming. Louis, to any ear except a Frenchman's, perhaps, didn't have a sufficiently celestial ring to it. Not until the mid 1800s, after a suggestion by English astronomer John Herschel, did astronomers adopt mythological names for all these moons.

John Herschel's father, William, was the most distinguished British astronomer of his time—perhaps of all time—widely hailed for his discovery of Uranus but also the center of considerable controversy. The elder Herschel began his career not as an astronomer, but as a musician. Born in Hanover, Germany, on November 15, 1738, Friedrich Wilhelm Herschel, like Galileo, was the son of a professional musician, an oboist for the Hanoverian Guards. Friedrich Wilhelm followed his father's example and, at the young age of 14, joined his father's military band as a violinist and oboist. But the Seven Years' War cut his career short. At the Battle of Hastenbeck in 1757, the Hanoverian Guards were severely battered, and young Herschel, along with his older brother Jacob (also a musician), decided that a military career was just too dangerous. They relocated to England, where William (as he called himself in England) believed he could devote himself to peaceful pursuits.

It was a struggle at first. Herschel eked out a living copying scores, giving music lessons, and playing in occasional concerts, but eventually he began to prosper. A more or less steady income gave him time to compose music of his own (still occasionally performed today), and in 1762, he began to send money home to his brothers and sisters. Around this time, British churches were introducing organ music into their services, and new positions as liturgical music directors were beginning to open up. Though Herschel was not a trained organist, he began to practice and did well enough so that, in 1766, when a job opened up at the Octagon Chapel in Bath, Herschel was appointed to the post of organ master and choir director. He got on well with the members of the local community and was known for his energy and enthusiasm.

There was enough money from that job and from the lessons he gave to local students for Herschel to settle down. In 1770, he moved into a modest row house at 7 New King Street in Bath, a short walk from the famous mineral springs and just down the hill from the stately Georgian mansions where the wealthy gentry of London came to enjoy the country air and take the waters. William Herschel had become a prosperous member of British society, but he still kept in touch with his family, both through regular correspondence and occasional visits from his brothers. His younger brother Alexander eventually came to live with him.

It was his sister Caroline's plight, however, that concerned Herschel most. After their father's death, Caroline had dutifully assumed the role of housekeeper for her aging mother, but William knew that his sister was as musically talented and intelligent as any of his brothers and that she chafed under the drudgery of domestic life. In 1772, he managed to persuade his mother to let Caroline move to Bath (he offered to send his mother money to hire a servant to replace Caroline), and his sister moved into the top floor of the New King Street house. Though she willingly took on the responsibilities of managing William's house, she also was eager to see if she could make a career as a singer, with the support and encouragement of her brother, who gave her voice lessons and a seat in the Chapel choir. But circumstances were to soon intervene that would change the course of both the Herschels' lives—William Herschel was about to become an astronomer.

It was, at first, merely a diversion from the day-to-day demands of musical directing, teaching, and composing. Herschel would read books on mathematics, optics, and astronomy in the evenings simply as a way of relaxing by doing something different. Just as he had with music, he went at his new hobby with frenetic energy, scribbling equations and diagrams as he went along. His astronomical readings inspired him to try his hand at telescopic observing, and so he purchased materials to construct a telescope of his own. "I was so delighted with the subject," he wrote to a friend, "that I wished to see the heavens and Planets with my own eyes thro' one of those instruments."

Herschel's first optical efforts, refracting telescopes, which used lenses to collect and focus light, were a disappointment. As we noted earlier, telescopes of this design, though they had been adequate for Galileo's discoveries, were difficult to use and often unwieldy. But by the late 1700s, an alternate design for telescopes that used mirrors was coming into fashion. Isaac Newton had presented a plan for a reflecting telescope to the Royal Society of London in the 1670s, and astronomers had made major improvements to the basic design since then. Reflecting telescopes had the advantage of being shorter and easier to handle than refractors. Their mirrors could be made bigger than lenses, which meant they could collect more light and thus see fainter objects. And, because mirrors reflected all colors of light equally, they did not suffer from the prismatic effect of light passing through glass, which made star images seen through refracting telescopes look like they were surrounded by colored haloes. The only major drawback to reflectors was that the mirrors of the time were made of polished disks of speculum metal, an alloy of copper, tin, and other elements. If the disks were polished to a high sheen, they collected light very effectively, but the speculum surface tarnished quickly, so that reflecting telescopes frequently had to be disassembled and the mirror surface repolished. Active telescope owners often kept several mirrors on hand for rapid replacement.

So Herschel decided to get some speculum metal, grind himself a mirror, and construct a reflecting telescope of his own. According to an entry in his journal for 1773, he bought the entire metal stock and tools of a local mirror maker who was going out of business and arranged for a short apprenticeship to boot. "It was agreed that he should also show me the manner in which he had proceeded with grinding and polishing his mirrors, and going to work with these tools I found no difficulty to do in a few days all what he could show me, his knowledge indeed being very considerable."

In a short while Herschel's knowledge of astronomy became very considerable, and he was able to construct telescopes as good as any that were available at the time. He and Alexander turned most of the house into a workshop; the kitchen became a foundry, and metalworking and woodworking tools lay strewn about, along with all sorts of astronomical papers and implements. He began to observe the heavens at night, carefully jotting down what he saw, and Caroline was frequently conscripted to help out with the record keeping. Given the notoriously bad English weather, he treasured every clear night, and would occasionally absent himself suddenly from a social occasion when a break in the clouds appeared. Herschel's musical endeavors began to suffer. A music student at the time recalled a lesson with Herschel in a room "heaped up with globes, maps, telescopes, reflectors, &c., under which his piano was hid."

At the same time, though, Herschel's reputation as an observer and telescope maker was growing. Amateur astronomers began to stop by the Herschel house to stargaze or to chat about the heavens. One evening in December 1779, when Herschel was carrying his telescope out the front door to look at the Moon, a passerby stopped and asked for a look through the eyepiece. The next morning, the stranger showed up again at the Herschel house and introduced himself as Dr. William Watson, a member of the Royal Society in London, the premier scientific organization of the day. Watson introduced Herschel to other scientific luminaries of the time, and soon, Herschel was sending letters that were read officially at meetings of the Royal Society. With the aid of Caroline and Alexander, he was really pursuing two careers at the same time—a musician by day and an observer by night.

Herschel's nighttime work, we should note, was not just random stargazing, and his astronomical goal was not just to produce telescopes that saw more clearly than anyone else's. He believed that careful observation could reveal hidden truths about the structure of the universe, and he realized that such observations required systematic record keeping and precise analysis. His observations of the Moon, for instance, were aimed at measuring the heights of lunar mountains. An even more ambitious project, for which he solicited the help of Caroline, was a complete cataloging of the position and appearance of all the bright stars visible from England. Herschel called it a "review" of the sky. Its ultimate aim was to find double stars, pairs of stars which appeared close in the sky. Herschel believed these were just chance alignments and that one star would invariably be much closer than the other.

By measuring the separations between the stars in a pair, Herschel believed, he would eventually be able to detect the "annual parallax," a regular shift in the positions of nearer stars caused by the Earth's motion around the Sun. Measuring parallax was a holy grail to astronomers. It would provide direct confirmation that it was the Earth, not the Sun, that moved, and it would make possible the determination of the distance of the stars. Ultimately, Herschel was frustrated in this effort: the annual parallax is so small that it would not be detected until 1838 by a German astronomer, Friedrich Wilhelm Bessel. To further complicate matters, as Herschel later realized, almost all of the double stars he observed were real pairs, not accidental alignments. Each member of a pair, like a planet in a solar system, was closely bound by gravity to the other star and orbited around it. Thus, the annual parallax could not be seen; what he observed was the motion of each star in its orbit around the other.

Herschel, unaware of the underlying difficulty of the parallax project, continued his cataloging efforts. On March 13, 1781, he was alone in the back garden of a house he had recently moved into at 17 New King Street, carrying on his observations as always; Caroline, who might normally have been with him, was back at their previous residence wrapping up some business. Sometime between 10 and 11 in the evening, as his journal records, Herschel caught sight of an odd object. It was a fuzzy disk, not a pinpoint of light like a star, and his first impression was that it was "a curious either nebulous star or perhaps a comet." He watched carefully for several nights, and it was clear that the object was moving against the background of stars, so it was surely an object in the solar system.

Was it indeed a comet? As word of Herschel's discovery began to circulate, opinions varied. Nevil Maskelyne, the Astronomer Royal, wrote to Herschel's friend Watson, "I was enabled last night to discern a motion . . . which, as well as from its agreeing with the position pointed out by him, convinces me it is a comet or a new planet, but very different from any comet I ever read any description of or saw." Charles Messier, a noted French astronomer, wrote from Paris, "I am constantly astonished at this comet, as it does not resemble any one of those I have observed, whose number is eighteen." Still, Herschel was willing to believe it was some new form of comet, unless compelling evidence indicated otherwise. No one, you must remember, had ever discovered a new planet before, at least since the day the first cavemen noticed that some of the stars in the sky moved from night to night.

So on April 26, when Dr. Watson read a letter from Herschel to the Royal Society describing the new object, it was titled simply "Account of a Comet." Though Herschel had reported on his work to the Society in the past, he was still a relatively unknown correspondent, and, though the report stirred interest, there was also some dissent. Herschel's object seemed so strange, and in his letter he claimed to have observed it at magnifications of over 1,000 times. A magnification of 250, at that time, was considered about the limit for a good telescope, so Herschel's remarks seemed like the boasts of an unsophisticated zealot. Neither Herschel nor anyone else knew at the time, however, that his telescopes were, in fact, far superior to any in England, even those at the Royal Observatory, and his claims were indeed quite accurate. When Herschel had an opportunity, a year or so later, to conduct side-by-side comparisons, the excellence of his work became clear to all. "I can now say," he noted in a letter to his brother Alexander following these field tests of his instruments, "that I absolutely have the best telescopes that were ever made."

Herschel left it to the professionals to follow up on calculating an orbit for his comet. After presenting his work to the Royal Society, he continued on with his observations and his telescope making, viewing the comet as a minor diversion. Others, however, recognized that Herschel had discovered something remarkable, whatever it was. In November 1781, he was called to London to personally accept the Copley Medal of the Royal Society. Herschel took the night coach to the ceremony and returned to Bath as quickly as possible; there was observing to be done. A month later, he was voted a Fellow of the Royal Society—a distinct honor for a small-town organist.

By the time of the Copley award, it was becoming clear that Herschel's object was not a comet. A French mathematician, Pierre de Laplace, and a Swedish astronomer, Anders Johan Lexell, independently calculated the path of the new object from the accumulating positional measurements that Herschel and others had reported. Comets move in highly elliptical orbits, spending most of their time far from the Sun and then quickly looping in to the inner solar system, often crossing the orbits of the Earth, Venus, and Mars. But Herschel's object was in a more circular orbit, staying far from the Sun—almost twice as far from the Sun, in fact, than Saturn. This was clearly a planet—the farthest planet in the solar system. Herschel had not only been voted a Fellow of the select Royal Society, he now belonged to a distinguished group of astronomers of which he was the sole member—the first person in history to discover a planet.

There was some grumbling, even after Herschel's accomplishment was acknowledged, that he hardly deserved credit for just being in the right place at the right time. The discovery of the planet was a lucky accident, the argument went, and anyone who just happened to be looking through a telescope could have seen it. "This is an evident mistake," Herschel countered. "In the regular manner I examined every star of the heavens, not only of that magnitude, but many inferior, it was that night *its turn* to be discovered. I had gradually perused the great Volume of the Author of Nature and was now come to the page which contained a seventh Planet. Had business prevented me that evening, I must have found it the next, and the goodness of my telescope was such that I perceived its visible planetary disc as soon as I looked at it."

It mattered not what the cynics thought; Herschel now had friends in high places. Word came to Bath that King George III, who maintained a keen interest in science, was interested in rewarding Herschel for his accomplishment. Herschel's admirers, including Dr. Watson and Nevil Maskelyne, realized that Herschel's astronomical gifts could be put to better use if he were not burdened with the daily tasks of a musician. They lobbied King George to grant him royal patronage to pursue astronomy full time, just as Galileo had received the support of Cosimo de' Medici. In May 1782, George called Herschel to London to meet and discuss astronomy, and over the next several months, in a series of negotiations, Herschel and the king came to an agreement. Herschel would move to Windsor, close to the king's country residence. He'd receive a stipend of £200 a year, freeing him from his musical obligations. In return, Herschel would simply be called upon to hold stargazing sessions for the royal family when requested.

George III was not only delighted with Herschel's discovery, he was also so taken with the excellence of Herschel's optical craftsmanship that he ordered five telescopes be made for him. Herschel, as part of the patronage deal, was permitted to manufacture and sell his own telescopes, which he did with great success. There are many of Herschel's creations scattered around the United Kingdom today in museums such as the huge London Science Museum in South Kensington (which has a large display of George III's scientific instruments) and smaller museums in Cambridge and Oxford. (In 1795, long before the end of his career, Herschel noted that he had already ground 430 mirrors.)

Thus, in July 1782, Herschel and his family left Bath and moved to a house near Datchet, a small town near Windsor Castle. The building was ramshackle and rambling, but there was plenty of room for Herschel to work and ample space for setting up telescopes outside. There he began his second career as an astronomer, busily building telescopes and charting the heavens, while Caroline fretted about the upkeep of the old building and the maintenance of their nightly observing records. Herschel moved several times thereafter, always close to Windsor, and continued to observe productively until shortly before his death in 1822 when ill health made it impossible for him to get out at night. He married a widow named Mary Pitt in 1788, who gave birth to their son John in 1792. Though the first year after this was rocky for Caroline (she later burned all her personal papers from this period so that her dark thoughts on her brother's marriage might not be made public), she and her brother reestablished a comfortable relationship, and the two continued to collaborate in astronomical research. Caroline, who also received a royal subsidy to do astronomy, died in 1848 at age 98, honored not only as the sister of a great astronomer, but as an accomplished scientist herself. She was credited with the discovery of eight comets.

THE NAME GAME AGAIN

The details of Herschel's later work are outside the scope of a short book, since, shortly after the discovery of his planet, Herschel's interest returned to the universe at large. He didn't abandon the solar system entirely; as he built new telescopes, he often tested them out by pointing them at planets, and he had bursts of activity when he saw something interesting. In 1787, he discovered the two largest moons of his new planet and, in 1789, two smaller moons of Saturn. He continued to observe the new planet for years, hoping to find more satellites, and in 1798, he reported discovering what he thought were four faint moons in orbit about it. These, however, turned out to be ephemeral—they were background stars or glints of stray light in this telescope, not moons, and later observers failed to find them again. False sightings like this were not uncommon, as we shall see in later chapters, though Herschel was usually among the most careful of observers. (Recall that Galileo's observations of Jupiter's moons were at first questioned by some as being internal reflections in the telescope, a not unreasonable objection. One must always be careful when looking for faint objects like moons right next to very bright objects like planets.)

Herschel seemed almost relieved to leave the planetary system behind, reaching for the stars. Just a year after the discovery of the new planet, he wrote to Caroline, "Among opticians and astronomers nothing now is talked of but what they call my Great discoveries. Alas! This shows how far they are behind, when such trifles as I have seen and done are called great. Let me but get at it again!"

However, before he could turn to other things, Herschel faced the problem of what to name his planet. Sir Joseph Banks, the president of the Royal Society, regarded this as a matter of national importance. In November 1791, in a letter informing Herschel of the Copley Medal, Banks wrote: "Some of our astronomers here incline to the opinion that it is a planet and not a comet; if you are of that opinion, it should forthwith be provided with a name." The danger of delay, Banks warned, was that "our nimble neighbors, the French, will certainly save us the trouble of Baptizing it." (As we shall see in Chapter 5, that same British-French rivalry was to play a crucial role in the controversy surrounding the discovery of the next major planet, Neptune.)

Herschel was willing to oblige. To seal the deal with King George, and simply to honor his monarch, he thought it fitting to name the new planet *Georgium Sidus*, Latin for "George's Star." In a letter to the Royal Society in 1783, published in the Society's *Philosophical Transactions*, he wrote:

> In the fabulous ages of ancient times the appellations of Mercury, Venus, Mars, Jupiter, and Saturn were given to the planets, as being the names of their principal heroes and divinities. In the present more philosophical era, it would hardly be allowable to have recourse to the same method, and call on Juno, Apollo, Pallas, or Minerva, for a name to our new heavenly body. The first consideration in any particular event, or remarkable incident, seems to be its chronology: if in any future age it should be asked, when this last-found planet was discovered? It would be a very satisfactory answer to say, "In the reign of King George the Third." As a philosopher then, the name of Georgium Sidus presents itself to me, as an appellation which will conveniently convey the information of the time and country where and when it was brought to view.

You will find the seventh planet of the solar system referred to as *Georgium Sidus* in a few textbooks published in England in the late 1700s and early 1800s. The British *Nautical Almanac* referred to it as the Georgian until 1850. But, as had been the case with Galileo's Medicean Stars, no one outside the country rushed to adopt the name. Some simply called it "Herschel" or "Herschel's Planet"—a common label in English-language textbooks until the middle of the 1800s.

But the Olympian tradition, standard for the other planets, was immediately appealing to most Europeans. As early as 1783, Johann Bode, a noted Prussian astronomer (who we will meet shortly in the story of the asteroids), sent a letter to Herschel congratulating him on his discovery. "You know perhaps, that I was the first person in Germany to see the new star . . . ," he wrote. "I have proposed the name of Uranus for the new planet as I thought that we had better stick to mythology and for several other reasons. . . . I have further proposed the sign ⛢ for our new planet." In Greek myth, Uranus was Saturn's father, an apt name for the next planet beyond. And Uranus, in the end, is what people came to call it.

When we look back at the accomplishments of the great astronomers, Galileo stands out as a pioneering figure, the first to systematically employ the telescope for scientific discovery. Herschel, following in Galileo's footsteps, created optical telescopes of surpassing quality, without which he might not have been able to recognize Uranus as a planet. Yet despite, or perhaps because of, the novelty and importance of their achievements, neither could steer clear of controversy. Whether it was the discovery of the first moons of a planet beyond the Earth or the discovery of the first new planet beyond Saturn, astronomers maintained a tradition of wrangling about what was found, how it should be named, and who deserved credit for it. That contentious spirit, as we shall see in the following chapters, continues undiminished to this very day.

CHAPTER 4

Ceres and Theories: The Search for Planets Begins

Then felt I like some watcher of the skies
When a new planet swims into his ken

—John Keats, "On First Looking into
Chapman's Homer," 1816

Finding planets was all the rage following the discovery of Uranus. To be sure, there were a few who pooh-poohed the significance of Herschel's discovery—German philosopher Georg Christoph Lichtenberg quipped, "To invent an infallible remedy against toothache, which would take it away in a moment, might be as valuable and more than to discover a new planet"—but the vast majority of people, like English poet John Keats, admired and marveled at the unprecedented achievement. In one fortuitous leap, the span of the solar system had doubled, the number of planets had gone from six to seven, and the domain of human knowledge had taken a giant step outward toward the stars. Educated Europeans sensed that they lived in auspicious times. Just as the great explorers had gone out in frail ships during the 1400s and 1500s to discover new continents previously unknown to Western civilization, so the astronomers of the 1700s, equipped only with ingeniously crafted telescopes, would discover worlds undreamed of before. Why should the roster of the solar system stop at seven? Perhaps there were dozens of planets out there just waiting to be spotted.

Nevertheless, finding a planet beyond Uranus, astronomers realized, would not be a simple task. Had Herschel's telescope not been so exquisitely crafted, he might not have been able to distinguish the disk of a new planet from the dot of a star, and Uranus might have slipped into his star catalog unnoticed. Indeed, at least a score of previous observers had observed Uranus before Herschel without noting anything distinctive about it. John Flamsteed, the first Astronomer Royal of England, had recorded it a half dozen times, the first in 1690, each time failing to note both its small disk or its slow motion; it appears as an anonymous faint star in his observation logs. A planet beyond Uranus would likely appear smaller, slower, and fainter yet, making its discovery even more challenging.

There was, however, a good possibility that there were planets *closer* than Uranus that had not been discovered yet, especially in the seemingly empty gap between Mars and Jupiter. The gap not only seemed unusually large (Mars is about 150 million miles from the Sun and Jupiter about 500 million miles), but the natural balance of the solar system seemed to require a planet there.

A mathematics professor at the University of Wittenberg in Germany, Johann Daniel Titius, had first called attention to this peculiar state of affairs in 1766. Titius noted that if we expressed the distances of the planets from the Sun in astronomical units (the average distance between the Earth and the Sun), they seemed to follow a regular numerical sequence (though we should emphasize that neither Titius nor anyone since has been able to explain *why* the planetary spacings seem to follow this rule). Here's what Titius noted. Take the number 0, add 3, and then continue to double each successive number: 6, 12, 24, 48, 96, and so on. Add 4 to each number, yielding 4, 7, 10, 16, 28, 52, 100, and so on. Then divide each number by 10, and you get the sequence 0.4, 0.7, 1.0, 1.6, 2.8, 5.2, 10.0. These numbers are a pretty close approximation, with one exception, of the distances of all the known planets from the Sun. The following table shows the suggestive agreement.

Planet	Distance Predicted by Mathematical Sequence	Known Distance in Astronomical Units
Mercury	0.4	0.39
Venus	0.7	0.72
Earth	1.0	1.00
Mars	1.6	1.52
????	2.8	????
Jupiter	5.2	5.20
Saturn	10.0	9.54

The obvious discrepancy with the relation is at 2.8 astronomical units from the Sun, a breakdown of the rule that Titius could not help but notice. Though he was just guessing, to be sure, he was of the opinion that the gap was occupied by previously undetected satellites of Mars and Jupiter.

Titius might have been forgotten had he not drawn the attention of Johann Bode, a young German scientist with a growing reputation as a writer and astronomical researcher. In the 1772 edition of Bode's popular textbook, he noted the apparent mathematical law of planetary spacings, without giving Titius credit, remarking, "Can one believe that the Founder of the Universe left this space empty? Certainly not." Bode proposed something more substantial than satellites, however: he believed that there was a planet between Mars and Jupiter just waiting to be discovered. In later books, he acknowledged the work of Titius, and though the sequence of planetary spacings is often referred to simply as Bode's law, scholars today call it the Titius-Bode relation.

Coming as it did less than a decade after the first appearance of the Titius-Bode relation, Herschel's discovery of Uranus prompted a renewed interest in finding planets between Mars and Jupiter. Why should there not be a planet there when, at the outer edges of the solar system, Uranus was right where the mathematics said it should be? The next number in the sequence (try it yourself) is 19.6 astronomical units from the Sun, quite close to the distance that astronomers were deriving from the growing body of observations of Herschel's planet. So if Uranus was mathematically "expected," albeit only in retrospect, then perhaps there really *was* a missing planet in the Mars-Jupiter gap.

The problem was finding it. That it had escaped notice so far meant that the undiscovered planet was rather faint (the implication being either that it was small or that its surface did not reflect much light). Finding a dim, slow-moving object among the host of stars in the sky was even more difficult than finding a needle in a haystack. To succeed, one would presumably need to see a faint moving dot, and that would require examining the sky in detail, carefully cataloging every object one saw and looking on successive nights for objects whose position had changed. One could identify these objects by matching observations against sketches made on previous nights. In an age when all such observations had to be made by hand (no photography or digital imaging existed, of course), this was a daunting task and certainly not one that could be carried out by a single astronomer.

Hungarian astronomer Franz Xaver von Zach began thinking about these difficulties following a meeting with William Herschel in England in 1783, where the two had discussed the remarkable agreement of the Englishman's discovery with the predictions of Bode's formula. Von Zach took up the matter of the missing planet with Bode himself when he returned to the continent and made a few rough calculations on where to look between Mars and Jupiter. A few years later, von Zach became director of a new observatory in Gotha, Germany, and continued to discuss the missing object with Bode and other associates. He felt a growing conviction that the missing planet was just waiting to be found.

Not everyone, however, agreed. Jerome Lalande, a preeminent French astronomer, who visited von Zach in Gotha in 1798, downplayed the significance of Bode's law. Its predictions, though close, were not in exact agreement with observations, Lalande objected, and in the absence of a theory of why the law took the particular form it did, it was likely that the approximate agreement was just a coincidence. Don't put your faith in number trickery, Lalande counseled, for you have no proof that there is any substance to Bode's law.

The proof of Bode's law, von Zach felt, lay in putting it to the test of finding the missing planet, no matter how difficult that might be. To carry out a systematic search for a planet, it was clear, he'd need help; a coordinated international effort was essential. Accordingly, on September 21, 1800, von Zach and Bode assembled a group of six astronomers at the home of Johann Schroeter in Lilienthal, a small town not far from Bremen in northern Germany. They decided to organize a search, dividing the zodiac up into 24 parts and assigning one section to each of 24 observers around Europe, astronomers known for their skill at the eyepiece of a telescope. The astronomers included Bode and von Zach, of course, and also such luminaries as William Herschel at Windsor, Nevil Maskelyne in Greenwich, and Charles Messier in Paris. They joked that the band of 24 should be considered "Celestial Police," since their goal was to enforce Bode's law. Von Zach, as secretary of the organization, sent out letters inviting each of the 24 to begin observing, informing each of his allotted portion of the zodiac.

One of the designated Celestial Policeman was Father Giuseppe Piazzi, a 54-year-old monk from Sicily, who had a reputation as a cautious and assiduous observer. Following his studies in Italy and France as a young man, Piazzi taught philosophy, science, and theology in Malta and at various Italian schools, and since 1780, he had been a professor of mathematics at the University of Palermo. In the late 1780s, when a new observatory was being built at the viceroy's palace in Palermo, Piazzi was asked to take over its operation. Though previously untrained in the subject, he schooled himself in the technical details and traveled to England to buy a quality telescope for his observations. He went to Greenwich for consultations with the Astronomer Royal, Nevil Maskelyne, and also visited William Herschel, where, one night at Herschel's large telescope, he reportedly fell off the observing platform and fractured an arm.

Undaunted by this harsh introduction to practical astronomy, Piazzi returned to Sicily, and in 1791, when the Palermo Observatory was completed, he began nighttime observing, taking it upon himself to compile a catalog of star positions, rather like the all-sky survey that William Herschel was conducting. It required him to be at the telescope every clear night, meticulously measuring star positions through the eyepiece.

On New Year's Eve before January 1, 1801, Celestial Policeman Piazzi hit pay dirt, though at the time he did not even know he had been deputized by Bode and von Zach. (The mails were slow, and the letter from Lilienthal did not arrive until some weeks later.) While his neighbors were welcoming the new year with pasta and wine, he was at the telescope squinting through the eyepiece and writing down coordinates. Early in the evening, he noted a star that was not in the catalog at all; the next night the star had moved, and it soon was clear that it was a small object in orbit around the Sun. Still unaware of his involvement in the Lilienthal search, he sent descriptions and coordinates to Bode and to several other astronomers. He wrote to a friend in Milan, Barnaba Oriani, at the end of January: "I have announced this star as a comet; but the fact that the star is not accompanied by any nebulosity and that its movement is very slow and rather uniform, has caused me many time to seriously consider that perhaps it might be something better than a comet."

It was, indeed, something better, though Piazzi was cautious about concluding that it was a planet. For a while, in fact, it seemed that his discovery might slip away from him before its true nature could be determined. During the entire month of January, Piazzi was the only astronomer who knew about the new object; Bode did not receive his letter until March 20, and Oriani didn't receive his until April 5. Worse still, in early February, Piazzi took ill and was unable to continue tracking his discovery. By the time he recovered, the object was no longer visible at night. It had orbited too close to the Sun and would not be well suited for observation for another few months.

In effect, Piazzi had lost the object, because its track was not yet well determined enough to know just where it would emerge from conjunction with the Sun, and it was so faint that it could not easily be spotted without knowing exactly where to look. When Bode finally received word of Piazzi's discovery, he must have regarded it with bitter irony. The planet he was fishing for had been caught, but before he had had a chance to inspect it himself, it had slipped the hook and swum away.

Or so it seemed. Enter Carl Friedrich Gauss, one of the greatest of all mathematicians, who was then a young scholar as yet unsure of his true calling, making a living by teaching classes in Brunswick, Germany. The problem of the planet that got away fascinated him as soon as he heard about it. Piazzi had only observed the object for 41 days, and no one had ever before calculated the orbit of a planet from such meager observations. The mathematical challenge this posed, and the chance to get involved in the great game of planet finding, appealed to Gauss so much that, as historian Agnes Clerke wrote, "the news from Palermo may be said to have converted him from an arithmetician into an astronomer." Gauss devised an elegant method to compute the orbits of planets using only a few observations. (Known as the "method of least squares," it is used today in a wide variety of applications—it's part of the program in your car's GPS unit that turns the signals from orbiting satellites into precise values of your latitude and longitude.)

Using Piazzi's observations of the missing object, Gauss was able to predict where to look for it when it became visible at night once again. He sent his tables to the Celestial Police in November, and the members of the Celestial Police force went scurrying to their telescopes. Frustratingly, the autumn of 1801 was a soggy one in Europe. For the first few weeks after Gauss's positions arrived, searchers saw mostly mist, drizzle, and sleet, with only a few glimpses of clear sky through breaks in the clouds. Finally, on December 31, 1801, just one year after Piazzi had first spotted it, von Zach saw the much-awaited object almost exactly where Gauss said it would be. Two nights later, another member of the Lilienthal group, German physician Wilhelm Olbers, a devoted amateur astronomer, saw it, too, from his observatory in Bremen.

From then on, astronomers around Europe never lost track of it. As Piazzi's curious object moved through the constellations, it became clear that its orbit was that of a planet, located just where the Titius-Bode law predicted. Any disappointment that the Celestial Police might have felt at being upstaged by an accidental discovery made at a provincial observatory even before the systematic search had begun was tempered by their elation that the mathematical harmony of the solar system had so strikingly been validated. Bode gloated over the "confirmation of that beautiful progression," which, in his opinion, put to rest any objections that skeptical astronomers may have raised about numerical coincidences. Bode's law—whatever the physical process that produced it—was more than just a game of numbers. After all, this was the second time (if we count Uranus) that the curious mathematical relation had come through with a right-on-the-money prediction.

Bode's satisfaction was made even sweeter by the publication, early in 1801, of *Philosophical Dissertation on the Orbits of the Planets* by a young philosopher from Jena named Georg Wilhelm Friedrich Hegel, one of the founding fathers of German idealist philosophy and the formulator of a method of "dialectic" analysis that was to influence Karl Marx a half century later. Hegel, unaware that Piazzi had already found a planet between Mars and Jupiter, argued in *Dissertation* that the logic of Bode's law was flawed and that it could not be used to predict "missing" planets. He wrote:

> While the displacements of the planets suggest an arithmetic progression in which, unfortunately, no planet in nature corresponds to the fifth member in the series. . . . Since this progression is arithmetic and does not follow a number series that generates them itself, i.e. not by powers, it is of no interest to philosophy.

Hegel's argument was a technical one, based on a close reading of Plato, but his bald statement "it is of no interest to philosophy" gave rise to the popular belief that he'd denied the existence of an eighth planet even as the news of that eighth planet was being made public. Whether or not that was his intention, Hegel's *Dissertation* was an embarrassment for a man so confident in the power of the human intellect—especially his own.

To add to these technical controversies, there was, as in the past, the issue of a suitable name. Arguably, naming new objects was more difficult than discovering them—at least judging by the experiences of Herschel and Galileo—so Piazzi was quick to assert his priority. "I have the full right to name it in the most convenient way to me. . . . I will always use the name Ceres Ferdinandea, nor by giving it another name will I suffer to be reproached for ingratitude towards Sicily and its king," he wrote. He chose Ceres for the patron goddess of Sicily, the classical goddess of the harvest, and Ferdinandea after Ferdinand I, the reigning monarch of Sicily. Not surprisingly, only the classical name stuck, and Ceres is what we call it today.

But was Ceres the planet that filled the gap? It was clearly smaller and fainter than expected, and it was possible that it was just a fellow traveler to a larger planet between Mars and Jupiter. On March 18, 1802, while searching for comets, Olbers spotted another moving dot of light that very much resembled Ceres. Like Ceres, it was faint, so small that no disk could be distinguished, and orbiting about 2.8 astronomical units from the Sun. Olbers proposed the name Pallas, after Pallas Athena, the Greek goddess of wisdom and of weaving. Neither Ceres nor Pallas seemed to fit the description of the missing planet, and the space between Mars and Jupiter was beginning to look a mite crowded.

Were these planets at all? William Herschel estimated that the two new objects were no more than 200 miles in diameter, several hundred times smaller than Uranus. Perhaps wishing to protect his status as the sole member of the planet-discoverer's club, Herschel proposed that Ceres and Pallas were the first of a new type of solar system object—not planets or moons or comets. Since they showed no apparent disk, he suggested the name *asteroids*, or "starlike objects." Piazzi, granting the unique nature of the new worlds, countered with *planetoids*, which had the virtue of avoiding the impression that these objects were stars. Two centuries later, the argument over how to classify the first asteroids was mirrored in the debate over the reclassification of Pluto.

Whatever the nature of these small bodies, it was beginning to look like there were a lot of them. On September 2, 1804, Carl Harding, an assistant of Schroeter's, found a third world, dubbed Juno, in the constellation Pisces, and on March 29, 1807, Olbers found another object, named Vesta, in the constellation Virgo. These, like Pallas, were smaller and fainter than Ceres. But what were astronomers to make of this plethora of discoveries? Instead of one gap-filling planet, there now were four, swelling the ranks of the solar system. In 1800, there had been 7 planets orbiting the Sun; by summer 1807, there were 11. Nothing in Bode's law had prepared astronomers for such an embarrassment of riches.

Fortunately for Bode and his contemporaries, the spate of discoveries stopped as quickly as it had begun. Though astronomers searched energetically, no new asteroids were discovered for almost 40 years. It was a measure of the difficulty of the task, for the excitement of finding new planets was a powerful motivation to proceed, and the Celestial Police continued their observations with vigor, joined by a cadre of others who hadn't been deputized by von Zach and Bode. The problem was not only that there was an awful lot of sky to search, but also that star charts in those days were inadequate to the task of recognizing faint objects.

During these decades, however, planet searchers began to accumulate observations and to create star charts that were more accurate, complete, and better tailored to the job of planet finding. Harding, most notably, compiled *Celestial Atlas*, which contained charts that represented what an observer would see through a telescope; previous celestial charts had been cluttered with artistic representations of the constellations that were more of a distraction than an aid to practicing astronomers. Though Piazzi died in 1824 without seeing any further discoveries, Harding in 1834, and Olbers in 1840, they and their colleagues laid the groundwork for future successful surveys, so that when asteroids finally began to be discovered again, they were discovered in great numbers.

The logjam broke in December 1845, when an amateur astronomer named Karl Ludwig Hencke—former postmaster from the Prussian town of Driesen—found the fifth asteroid, named Astraea. Over the next 4 years, the trickle of asteroids turned into a flood, with 10 more discoveries in rapid succession: Hebe (July 1847), Iris (August 1847), Flora (October 1847), Metis (April 1848), Hygiea (April 1849), Parthenope (May 1850), Victoria (September 1850), Egeria (November 1850), Irene (May 1851), and Eunomia (July 1851). Finding and naming asteroids became a growth industry, albeit one that required a lot of time and effort. In addition to finding the asteroid, an erstwhile discoverer had to record its position, night after night, for many months, so that one could calculate its position any time in the future. If such follow-up observations were not made, the asteroid could be "lost" as Ceres had been, and the discovery would count for little, since, after a while, no one would know where to look for the object.

By 1900, about 1,000 asteroids were on the books, and many astronomers, who were by then becoming interested in taking photographs of objects outside the solar system, were actually getting annoyed that asteroids seemed to be so ubiquitous. It was hard to avoid finding a trace of an unnamed asteroid on an exposure of the sky. When that happened, an astronomer would have the choice of dropping everything to observe the new asteroid—which was just one of thousands of presumably indistinguishable little chunks of rock—or ignoring it and going on with other work. Asteroids became known to astronomers as the "vermin of the skies."

None of this growing number of asteroids rivaled Ceres in size, and clearly none were the larger planet that the Celestial Police had been searching for. This caused a problem for astronomers and textbook writers, who were at first inclined to simply list the asteroids along with the planets. In the early 1800s, with only a few asteroids added to the seven big planets, no one gave much thought to semantically separating the asteroids from their larger brothers, but by the middle of the 1800s, the added load of asteroids was becoming unsustainable. In the 1849 edition of Asa Smith's *Smith's Illustrated Astronomy*, a popular textbook aimed at high-school students, the strain was beginning to show. It included the following entry (note the use of "Herschel" for Uranus and the presence of a still farther planet which had been discovered in 1846 and which we will discuss in the next chapter).

Q: How many primary planets are there?

A: Sixteen; eight being asteroids or small planets.

Q: What are their names, beginning with the sun?

A: Mercury, Venus, the Earth, Mars, (Vesta, Astraea, Juno, Ceres, Pallas, Hebe, Iris, Flora), Jupiter, Saturn, Herschel, or Uranus, and Leverrier, or Neptune.

In the 1850s, as the number of asteroids began to exceed the number of major planets, astronomers began to pay more attention to what distinguished the asteroids from the other planets, and the terminology in textbooks and professional publications began to reflect that distinction. Beginning in the 1860s, the asteroids were more and more frequently accorded separate status as a group of bodies in the solar system, and, in succeeding decades, astronomers started to use the term "minor planets" to refer to the bodies between Mars and Jupiter. The official office that keeps track of them all today (more than 400,000 as of 2009) is the Minor Planet Center of the International Astronomical Union.

Aside from the difficulty of keeping track of dozens of asteroid names, there was good reason for regarding the minor planets as a population distinct from "ordinary" planets. The obvious size difference was the most critical factor. Even though there was considerable uncertainty about their exact size, it was clear even in the 1850s that the largest asteroid, Ceres, was only a few hundred miles in diameter (the current value is about 560 miles), only a fifth that of the smallest planet, Mercury. Most of the asteroids were half that size or even smaller. It was becoming evident, in fact, that there were vast numbers of very small asteroids, perhaps only a few miles in size or less, that had not yet been spotted by astronomers because they were too faint to be seen with the telescopes available at the time. Clearly there had to be a reason that there were so many tiny rocky fragments orbiting the Sun only in this region. Surely there was something about the origin of the asteroids that had made them quite different from the major planets.

What was that something? At the time of his discovery of Pallas, Wilhelm Olbers advanced the idea that the asteroids were fragments of a full-sized planet that had somehow shattered, perhaps because of some internal instability or perhaps because it had collided with a comet. (No one knew at the time that comets, which are icy bodies only at most a dozen miles in diameter, were too small to have disrupted a planet.) As the number of known asteroids increased, the shattered-planet hypothesis at first looked like a plausible scenario. Yet if you sat down and worked out the math, the oddity of the asteroid belt between Mars and Jupiter didn't go away. Though the supposed original planet would, of course, have been larger than any of its fragments, it still would have been unusually small. Even today, when so many more hundreds of thousands of asteroids are known, the total estimated mass of asteroids between Mars and Jupiter is less than 5 percent of the mass of our Moon.

Astronomers currently view the asteroid belt not as the remains of a disintegrated planet, but as the debris of a failed planet. The modern view is that the planets aggregated themselves from bits of rock and ice that were orbiting the primordial Sun in a flattened cloud roughly 4.7 billion years ago. This so-called "solar nebula" was the cloud from which our Sun formed, and the planets were only leftovers that didn't get pulled by gravity into the Sun at the center. As these leftover pieces of material orbited, they would, from time to time, collide with each other and stick together, so that over time, like a snowball rolling down a hill, the smaller pieces would become larger pieces and the large pieces larger still. The biggest balls swept up material around them, and eventually the planets were formed, Jupiter being the largest of the group. The process of building planets by the accumulation of material from the solar nebula is known to astronomers as accretion.

Because Jupiter got there first with the most, its gravitational pull kept any substantial planet from forming between it and Mars. The gravitational tug of Jupiter tore apart growing bodies before they could snowball to planetary size. Jupiter also captured any stray bits of rock and ice that got too close and incorporated them into its already massive body, leaving very little material for an inner planet to build on. As a result, nothing of substantial size was able to form in the asteroid belt, and Ceres never grew even to the size of a small planet like Mercury.

Farther out from Jupiter, however, larger bodies could form without interference. That, of course, is why Saturn grew so big, as did Uranus. And that is why, as we shall see, an eighth planet formed another 10 astronomical units beyond Uranus. It remained, however, undetected by the Celestial Police and was not found until just after the discovery of Astraea, the fifth asteroid.

Looking back at the changing views of asteroids in the 19th century, we are struck by their similarity to the debates of the early 21st century about the nature of Pluto and the other bodies orbiting the Sun at the outer edges of the solar system. Pluto, as we shall see, was discovered in an attempt to satisfy a mathematical theory. Though it was initially hailed as a proof of the fruitfulness and correctness of the theory, Pluto turned out to be fainter and smaller than expected. As time went on, additional objects like Pluto were discovered, prompting astronomers to reconsider their decision to classify Pluto with the other planets. This led the International Astronomical Union to adopt a new category of dwarf planets. The 19th-century controversies about asteroids, in effect, were a rehearsal for the 21st-century reappraisal of our solar system that naturally came about as astronomers, object after object, compiled a complete inventory of the solar system.

CHAPTER 5

Neptune's Disputed Discovery

It is painful, that at the conclusion of a discovery which will ever be memorable—we find ourselves in the midst of broils and ungenerous jealousies and resentments—that constrain one to blush red on the mention of Man's unchallenged greatness!

—J. P. Nichol, *The Planet Neptune: An Exposition and History*, 1848

I was abused most savagely both by English and French.

—George Airy, *Autobiography of Sir George Biddell Airy*, 1896

While the Celestial Police manned their telescopes looking for objects closer than Jupiter, other astronomers were at their desks using pen and paper to search the solar system for planets farther than Uranus.

Not glamorous research, perhaps—though filling notebooks with calculations was no less appealing than shivering in an observatory dome all night. Yet the systematic search for undiscovered members of the solar system constituted what the Astronomer Royal of England, Sir George Biddell Airy, later called a "movement of the age" that would culminate in the prediction and consequent discovery of the planet Neptune.

To astronomers of the 19th century, the discovery of a planet beyond Uranus was regarded as the crowning triumph of celestial mechanics, the branch of theoretical physics that dealt with the motions of the planets. Many, in fact, regarded the discovery as the most important scientific achievement of the era. Astronomers believed they had learned how to read nature's coded message in the heavens, and a rising tide of optimism in the predictive power of physics inspired a host of later astronomical research, including, in the 20th century, the search that added Pluto to the roster of the solar system. Yet despite its high profile, or perhaps because of it, the discovery of Neptune ignited one of the fiercest scientific controversies of the century.

THE WANDERING WAYS OF URANUS

The astronomers who were looking for a trans-Uranian planet in the early 1800s were not observers but mathematicians, conversant in the abstruse calculus of celestial mechanics. Their principal aim, at least at the beginning, was to show that the behavior of the newly found planet Uranus could be mathematically accounted for by Newton's laws of motion. Using the law of universal gravitation and knowing the mass and position of the Sun and all the other planets, astronomers were able to compute the strength and direction of all known forces acting on Uranus at any point in time. Knowing those forces, in turn, enabled them to calculate the orbit of the planet and to compare it with observations past, present, and future.

The first computations went well. In the years immediately following Herschel's discovery, astronomers were able to predict with satisfactory accuracy where Uranus would be the next night, the next week, the next month, and the next year. But gradually Uranus began to stray from the predicted path. Whether something was wrong was at first unclear, because the observations were so limited in extent. For the purpose of comparing theory with observation, the planet moved at what seemed like a snail's pace. It took 84 years to go around the Sun, and any true test of whether a planet conformed to Newtonian theory required observations of the planet over several circuits around the Sun. With only a small fraction of its orbital arc in the record books, it was difficult to tell if Uranus was behaving precisely as expected.

Actually, there was a longer record of the motion of Uranus than astronomers thought at first. Though Herschel had been the first to recognize Uranus in 1781, astronomers soon found almost two dozen earlier sightings of the planet in catalogs and observing logs dating back as early as 1690. Over and over again, Uranus had been mistaken for an anonymous star, not because of carelessness on the part of astronomers, but because even the best telescopes in the days before Herschel did not have sufficient resolution to make its greenish disk stand out as something out of the ordinary. Through imperfect optics, Uranus looked like a featureless blob of light, just like a star seen through a typical telescope of the time.

Augmented by these prediscovery recorded positions, the orbit of Uranus could be traced through more than one complete orbit, which should have been sufficient to verify the effectiveness of Newton's laws. Yet to the dismay of astronomers, Uranus refused to cooperate: the new observations after 1781 and the old observations before 1781 could not be reconciled with each other, at least using the known laws of physics and the properties of the known planets. If you used the early observations to calculate its orbit, then Uranus should not have been exactly where Herschel discovered it in 1781. If you used the post-discovery data to calculate an orbit, then Uranus should not have been exactly where earlier observers recorded it.

French astronomer Alexis Bouvard, who investigated Uranus's orbit in the early 1820s, tried to remedy matters by questioning the reliability of the early observations and then disregarding them altogether. It was a conservative tactic for saving Newtonian physics, because the old observers were not around to defend the accuracy of their work. Bouvard was particularly harsh on his compatriot Pierre Charles Lemonnier, who had unknowingly recorded about a dozen observations of Uranus between 1765 and Herschel's discovery in 1781. Lemonnier was an innovative and brilliant astronomer who had conducted extensive observations of the Moon in the mid-1700s and who had even investigated how Saturn's gravitational force affected the orbit of Jupiter, an analogue to the problem astronomers were facing with Uranus. But Lemonnier was an easy target for criticism, because he was widely disliked in the French scientific community. Even two decades after his death in 1799, astronomers remembered his quick temper and the tenacity with which he held grudges. Once, because of some imagined insult, Lemonnier steadfastly refused to speak to a colleague for what he described as "an entire revolution of the moon's nodes"—that is, about 19 years!

Though Lemonnier's work could be called into question simply because he was a crusty character, it was not so easy to disregard sightings of Uranus by less controversial figures. The earliest sighting, in fact, had been made by England's first Astronomer Royal, John Flamsteed, an observer who was obsessive in his attention to detail. Moreover, it soon became evident that Bouvard's solution to the Uranus problem was a stopgap measure at best: even if only the most reliable post-discovery observations were included, the observed orbit of Uranus became more and more difficult to reconcile with theory with each successive year. No matter what data you used to calculate it, Uranus doggedly insisted on moving away from its predicted track. If the observations were not to blame, then only two possibilities remained to explain what was going wrong: either the Newtonian theory of orbits was flawed or some unaccounted planet was causing Uranus to stray from its calculated orbit.

There was an outside chance that the laws of physics were at fault, but up to this time, Newton's theory had proven remarkably effective in predicting the motion of objects in the solar system. Its most notable success had been Edmond Halley's recognition that a comet seen in 1682 had previously been seen in 1531 and 1607, orbiting the Sun in a highly elongated path that returned it to prominence in the nighttime sky every 76 years or so. Halley's computations predicted a return of the comet in around the end 1758, and when a German farmer, Johann Georg Palitzsch, caught sight of it on Christmas night of that year, Halley's fame was assured. The astronomer, who had died sixteen years earlier, did not get to savor his success, but the great comet he identified still bears his name and still returns to excite public interest every three-quarters of a century. When Halley's comet returned again in 1835, astronomers awaited it with sublime confidence. The agreement between observation and prediction was far better than in 1758; improvements in mathematical techniques, it seemed, only served to confirm Newton's laws.

No, a more palatable explanation for the errant behavior of Uranus was the gravitational attraction of an as-yet-unknown object in the solar system that was causing it to deviate from its calculated path. If that unknown object could be identified, then the behavior of Uranus would no longer seem strange, and the old observations of its positions would automatically snap into agreement with the new ones. Rather than search wide swaths of the sky for such an object (as the Celestial Police were doing in their search for asteroids), it might even be possible to use the deviant behavior of Uranus to help restrict the area that needed to be observed. Perhaps the deviations of Uranus from its predicted orbit could be used to calculate the position of the planet causing those deviations, and astronomers could then point their telescopes right at it, rather than laboriously scouring the entire sky for a moving point of light.

The mathematics needed to resolve the issue, however, had not yet been devised. Newton, Halley, and those who followed had perfected techniques for calculating the orbit of a planet when the masses and orbits of all the other objects in the solar system were known. But finding an unknown planet from the deviant behavior of another one was what modern mathematicians might call an "inverse" problem: given the observed orbits of every other object in the solar system *except* one, could the properties of that one remaining object be deduced? It was not at all certain that a unique solution for the missing planet could be determined from the available information, but even if it could, it seemed likely that the task would not be easy. George Airy wrote in the 1840s, "[I]f it were certain that there were any extraneous action [an unaccounted-for force], I doubt much the possibility of determining the place of a planet which produced it. I am sure it could not be done till the nature of the irregularity was well determined from several successive revolutions [of Uranus]." In other words, several centuries of observations were needed.

Difficult as the task appeared to Airy, the existence of a trans-Uranian planet was widely accepted among astronomers by the 1840s, and the discovery of that planet was high on the agenda of pressing scientific problems. But, unless the planet were to swim into view by some lucky happenstance, like Uranus, the search for it would require supreme mathematical talent, the patience to undertake lengthy calculations, and a healthy measure of youthful audacity.

That was, perhaps, an apt description of John Couch Adams, an undergraduate at St. John's College, Cambridge, who decided to solve the Uranus problem after reading an 1832 report by George Airy on the state of research in astronomy. In July 1841, he jotted down on his "to-do" list the following entry:

> Formed a design, in the beginning of this week, of investigating, as soon as possible after taking my degree, the irregularities in the motion of Uranus, which are yet unaccounted for, in order to find whether they may be attributed to the action of an undiscovered planet beyond it, and, if possible, thence to determine approximately the elements of its orbit, etc., which would probably lead to its discovery.

ADAMS TO THE RESCUE

Born in 1819 on a tenant farm in Cornwall in the far southeast of England, John Couch Adams might have lived a life of obscurity had he not been so enamored of numbers. He easily absorbed every mathematics book he encountered, quickly outpacing the abilities of his first tutor and dazzling his later teachers with his grasp of abstract concepts. While barely a teenager, he taught himself calculus and astronomy and at age 20 won himself a scholarship to St. John's College. It was remarkable for a country boy to matriculate there and even more remarkable that Adams excelled in all his studies, impressing his classmates with his candor, his calm assurance, and his modest demeanor.

Adams was barely 22 when he read Airy's account of the Uranus problem, but he was clear-sighted enough to realize that if he were to devote sufficient energy to solving it, he would first have to complete his studies. In 1843, he was named "senior wrangler" in mathematics in recognition of his preeminent performance in the grueling examinations that qualified him for graduation from Cambridge. That same spring, he was elected a fellow of St. John's College, a position that afforded him a room, a place at the High Table meals, and the opportunity to support his own work while tutoring other Cambridge students. It was time to get to work on the mystery of what was pulling on Uranus.

There were so many possible configurations for a suitable planet that it was necessary for Adams to make some assumptions. For instance, a massive planet far from Uranus could produce the same effect as a smaller planet much closer in. So Adams used a bit of guesswork, assigning the unseen planet an orbit about 38 astronomical units from the Sun. This was about twice as far from the Sun as Uranus and in rough agreement with the Titius-Bode relation. Though no one knew why that "law" worked, it seemed to have been able to predict the positions of both Uranus and the first asteroids, so it was worth a try. Assuming this distance, Adams worked backwards from the observed positions of Uranus to calculate the size and direction of the force that was perturbing its orbit.

It didn't take Adams long to get results. By October of 1843, he had worked out a tentative solution that gave him confidence he was on the right track. But he needed to be sure he had the most accurate and complete data on the positions of Uranus to feed into his calculations. Accordingly, he sought help from James Challis, who held a chair in astronomy at Cambridge and was director of the Cambridge Observatory. In response to a letter from Challis in February 1844, George Airy sent back the latest compilation of Greenwich Observatory data on Uranus, along with an invitation for young Adams to write him a letter describing his work.

With the new data in hand, Adams began to fully realize the magnitude of the task he had undertaken. He worked on the computations for another year and a half, using both new and old observations of Uranus, until, in September of 1845, he arrived at what he thought might be a satisfactory solution for the orbit of the planet that was influencing Uranus. He sent his predictions to Challis, presuming that the observatory director would quickly mount a search for the new planet with the Cambridge telescope. Adams had no way of knowing that his brilliant work would not be matched by equal diligence on the part of Challis and other English astronomers. Nor did he know that, just across the English Channel, another brilliant mathematician was hot on the same trail.

THE FRENCH CONNECTION

Urbain Jean Joseph Le Verrier was eight years older than Adams and, by the mid-1840s, one of the leading lights of the French scientific establishment. Like Adams, he was outstanding at mathematics, graduating at the top of his class from the École Polytechnique in Paris and quickly establishing a reputation for brilliance and hard work. Unlike Adams, however, he made a name for himself in a variety of fields. His first work was in chemistry, and he only took up astronomy after a position in that field became available in 1837. Le Verrier taught himself enough to land the job, and soon he was publishing papers in celestial mechanics and calculating the orbit of the planet Mercury—work that we will discuss further in Chapter 6.

In 1845, at about the same time that Adams was wrapping up his computations, Le Verrier was approached by François Arago, the most well-known French astronomer, who encouraged his colleague to apply the same mathematical skills he had displayed in analyzing the orbit of Mercury to the more urgent problem of Uranus. Le Verrier took up the challenge with a vengeance, and by November, he shared his first results in a paper presented to the Paris Academy, showing that, at least in principle, the old and new observations of Uranus could not be reconciled unless there was an eighth planet in the solar system. Six months later, in June 1846, he reported back to the Academy on his further analysis, including a computation of the current position of the as-yet-undiscovered body.

Up until this time, neither Le Verrier nor Adams knew of the other's work. Yet, remarkably, both had come up with essentially the same results. Though they differed in some details, such as the exact mass and size of the planet, the position that Le Verrier predicted in 1846 was within one degree of the position predicted by Adams nine months earlier. And, in fact, both predictions were close to the actual position of the planet. By all rights, English astronomers should already have discovered the unknown object by the time Le Verrier reported his results. But, unfortunately, they hadn't.

SKEPTICISM AND DELAY

When he received Adams' prediction of the position of the supposed new planet in the fall of 1845, James Challis, who had himself been an outstanding mathematics student, was not spurred into immediate action. Though he had high regard for Adams' abilities, he still lacked confidence in the results. Perhaps he did not fully comprehend the urgency of the situation—after all, astronomers had been speculating about a planet beyond Uranus since the time of Herschel's discovery. "It was so novel a thing to undertake observations in reliance on merely theoretical deductions," Challis later said in his own defense, and "while much labour was certain, success appeared doubtful." Challis was also, by nature, too sanguine to get embroiled in a frenetic race to establish scientific priority, and his later work as an astronomer, though good, never seemed to fulfill the promise of his early academic achievement. The task at hand called for what a later astronomer called "a bolder and more energetic man." Rather than rush to the telescope, Challis recommended that Adams seek advice from George Airy, and he provided a letter of introduction to Adams so that he could meet personally with the Astronomer Royal during an upcoming visit to London.

Unfortunately, Airy was attending a meeting in Paris when Adams arrived at Greenwich in late September 1845. As one of the most prominent scientists in England at the time, Airy was involved in numerous government and commercial advisory committees (many having little to do with science), and this distraction, in hindsight, kept him from devoting appropriate attention to Adams' claims. Nevertheless, when he returned, Airy sent Challis a note that he regretted missing Adams and was very interested in any developments regarding his analysis of the orbit of Uranus and a possible new planet. In October, Adams visited Greenwich again but once again found Airy out of town, and when he returned a third time, in mid-afternoon, a butler told him that the Astronomer Royal was at dinner (he had a habit of eating early) and could not be disturbed. Adams had good cause to feel rebuffed, and it is certain that Airy, if not hostile, was at least indifferent. The Astronomer Royal, for all his talent, was a stiff and formal man who had a prim awareness of academic rank and social position: he was not about to fawn over a Cambridge mathematician who had no established credentials or publications simply on the say-so of James Challis.

Though frustrated in his attempt to see the Astronomer Royal, Adams left a note for Airy giving an overview of his methods and tables of his computations. To his credit, Airy shared the results with William Dawes, one of several wealthy amateur astronomers in England at the time who possessed sizeable telescopes and conducted first-rate observations. Dawes' observatory was still under construction, so he was unable to test Adams' predictions, but he sent the calculations on to someone who could: William Lassell, a Liverpool brewer whose personal reflector, 24 inches in diameter, was the largest telescope in England. Again, Adams' predictions might have been fruitful, but as Lassell later recalled, a sprained ankle kept him from spending time at the telescope, and during his recuperation he somehow misplaced the letter from Dawes. In any case, Adams' predictions of September 1845 were not followed up on by any British astronomers during the months immediately following his communication with Challis and Airy. Both Challis and Airy were to regret the oversight.

In addition to communicating with Dawes, George Airy also replied to Adams in a letter written several weeks after Adams' visit. In it, he seemed to have misunderstood the elaborate calculations involved in the analysis of Uranus's orbit. He suggested that Adams try using a smaller value for the distance between Uranus and the Sun to see if that would reduce the discrepancy between the observed and calculated position of Uranus, apparently unaware that Adams had been using this very discrepancy to deduce the position of the new planet. Irritated by the indifference he sensed in Airy's comments, and still smarting from the apparent rebuff during his visit to Greenwich, Adams was slow to answer, and so again the search was delayed. As far as Airy was concerned, after November 1845, the ball was in Adams' court, and so he did not press for further information on where to find the supposed planet.

The Astronomer Royal's attitude changed considerably in June 1846 when he received a copy of the paper Le Verrier had recently presented to the Paris Academy. While he had been leery of Adams' result, he regarded Le Verrier as a worthy gentleman whose professional bona fides were well established. Airy was clearly impressed that the Frenchman and the Englishman, working independently of each other, had made such similar predictions of where to look for the new planet. He later recalled, "[N]ow I felt no doubt of the accuracy of both calculations," and he sent a letter to Le Verrier expressing his appreciation of the work. But inexplicably, he neither communicated with Adams nor mounted an immediate telescopic search. Not until July, possibly as a result of a conversation with Dean George Peacock, the Lowndean Professor of Geometry and Astronomy at Cambridge, did Airy finally realize what was at stake. He penned a hasty note to Challis, asking him to undertake a systematic search for the phantom planet. "I am asking," said Airy, "almost at a venture, in the hope of rescuing the matter from a state which is . . . almost desperate . . . the time for the said examination is approaching near." Though initially startled and a bit upset by Airy's tone (after all, Airy had known about the supposed planet for almost nine months), Challis undertook the task and, by the end of July 1846, had begun to sweep the heavens. But he did not head straight for the target. Still unsure of the value of the theoretical predictions, he established a very large survey region (30 degrees wide and 10 degrees high) centered on the calculated positions of Adams and Le Verrier. Searching this region exhaustively, he knew, would take many nights—probably more than one season given the vagaries of English weather. Again, Challis's conservative approach was to work against his success.

Word of Le Verrier's prediction and of Challis's search was beginning to spread (though Adams was still relatively unknown in the scientific community), and there was a growing sense that the search for the long-anticipated planet was nearing its climax. Sir John Herschel, son of the discoverer of Uranus, gave a public address in Southampton on September 10, 1846, noting that, after a hiatus of almost four decades, a fifth asteroid (Astraea) had been discovered just a year earlier and that this discovery of a new planet, small though it was, had renewed interest in the "probable prospect of the discovery of another. We see it as Columbus saw America from the shores of Spain. Its movements have been felt, trembling along the far-reaching line of our analysis, with a certainty hardly inferior to that of ocular demonstration." The "trembling" Herschel alluded to was the irregular behavior of Uranus, which was pointing with "certainty" toward a major planet lurking at the outer edges of the solar system.

Yet even as Herschel spoke, and even while Challis's search was getting under way, events were occurring on the other side of the English Channel that would snatch priority away from British astronomers.

A DRAMATIC DISCOVERY

In Paris, Urbain Le Verrier had not been idle. On August 31, 1846, he presented a paper to the French Academy describing refinements in his calculations that enabled him to better define the orbit of the new planet. Not only could it be discovered by noting its motion across the sky, he predicted, but through a high-quality telescope, it should appear as a clearly distinguishable disk, slightly smaller than Uranus. By this time, Le Verrier believed he had enough information to make the discovery of the planet possible without a protracted search. But, like Adams, he had difficulty convincing astronomers in France of the urgency of undertaking immediate observations. On September 18, just a week after Herschel's confident pronouncement that a discovery was imminent, Le Verrier sent a letter to Johann Galle, an astronomer at the Berlin Observatory, suggesting that the young man take a look with the nine-inch refractor at his disposal. This time there was no delay; on the evening of the day that the letter arrived, September 23, 1846, Galle and an assistant, Heinrich d'Arrest, pointed the Berlin telescope toward the coordinates specified by Le Verrier and began to search.

They might have spent days scrutinizing the many stars in their field of view had they not had the benefit of a new star map of the area that had just been drawn up for the purpose of finding asteroids by Carl Bremiker, a mapmaker at the Berlin Observatory. With the new map in hand, the two astronomers were able to tick off stars one by one, Galle at the telescope and d'Arrest checking against the map. In less than an hour, they had found something—a star that was not on Bremiker's chart. Excitedly, they watched it all night, checking surrounding stars to make sure there was no mistake. By the next night, they were certain—the "star" had moved, quite clearly, and they were pretty sure that it had a disk, too. On the morning of September 25, Johann Encke, director of the Berlin Observatory, sent out an official announcement that a new planet had been discovered at the edge of the solar system. The same day, Galle sent what must have been a stunning response directly to Le Verrier (in those days before e-mail, this must have been the first acknowledgment that Galle had received Le Verrier's request for observations). "The planet whose position you have pointed out *actually exists*," he wrote. "On the day on which your letter reached me, I found a star of the eighth magnitude which was not recorded in the excellent map designed by Dr. Bermiker." Galle went on to propose that the new planet be called Janus, the Roman god of doorways, beginnings, and endings. Le Verrier responded with gratitude, suggesting, however (perhaps wishing to reclaim the discoverer's right to a name), that the new planet be called Neptune.

TRIUMPH AND RECRIMINATION

Word of the discovery of a new planet spread rapidly, not just among astronomers but throughout the public at large. The story of how a master mathematician had been able to discover a planet using pure powers of reason was immensely appealing, especially in an age enchanted with the notion of progress and the power of science to transform civilization. Le Verrier became the toast of Paris, hailed as a giant among intellectuals, a man who had, in the words of compatriot Camille Flammarion, "discovered a star with the tip of his pen, without other instrument than the strength of his calculations alone." Arago, as the unofficial spokesman for French science, proclaimed the discovery of Neptune to be the crowning achievement of applied mathematics and a lasting gift of French science to mankind. Scottish astronomer J. P. Nichol marveled, "The entire annals of Observation probably do not elsewhere exhibit so extraordinary a verification of any theoretical conjecture adventured on by the human spirit!"

News of Neptune's discovery reached England around October 1, 1846, when the *Times* in London described the German observations and noted that they had been confirmed by John Hind, an English amateur who had no trouble finding the object. The existence of a trans-Uranian planet was no longer a matter of speculation, but debate over who deserved credit for its discovery had only just begun.

Up to this time, astronomers on the Continent knew nothing of Adams' work. Imagine their shock and surprise, then, to read an article by John Herschel in the London magazine *Athenaeum* published on October 3. Herschel gave full credit to Le Verrier for his genius and his labor, but he revealed that Le Verrier had not been the only one to compute an accurate position of Neptune. "The remarkable calculations of M. Le Verrier . . . if uncorroborated by . . . independent investigation from another quarter—would hardly justify so strong an assurance," he wrote, referring to the assurance that the planet could be discovered by theoretical methods. "But it was known to me at the time . . . that . . . a similar investigation had been independently entered into, and a conclusion as to the situation of the new planet very nearly coincident with M. Le Verrier's arrived at (in entire ignorance of his conclusions), by a young Cambridge mathematician, Mr. Adams."

It may be that John Herschel only wanted to grant Adams his due, but the reaction to the news in Paris was immediate and bitter. To Frenchmen, it seemed as if the British were trying to steal a good measure of their glory and perhaps even snatch the claim of priority away from them entirely. Arago accused the English astronomers of exaggerating "the merit of the clandestine researches of M. Adams." He continued, "This pretension will not be admitted. . . . No. No, the friends of science will not permit the consummation of such crying injustice. *Mr. Adams has no right to figure in the history of the new planet, neither by a detailed citation, nor even by the slightest allusion.*" Le Verrier, though less rhetorical, was no less concerned about retaining priority for his triumph. He wrote privately to Airy, wondering: "Why has M. Adams kept silent for four months? Why had he not spoken since June if he had good reasons to give? Why wait till the planet has been seen in the telescope?"

Of course, Adams had *not* been waiting idly by while astronomers on the Continent had been searching the heavens; Challis, albeit tardily, had taken up the quest for Neptune over a month before its discovery in Berlin. Yet the tactics adopted by the Cambridge astronomer raised further controversy. By Challis's own account, once informed of the discovery of Neptune so close to Adams' predicted position, he went back to his own records and found to his dismay, that he'd actually observed the planet three times since July, each time failing to notice its motion. "After four days of observing, the planet was in my grasp if only I had examined or mapped the observations," he remarked soon after learning of Galle's discovery. Challis did not, granted, have the benefit of the excellent star map that Galle and d'Arrest used, and so he had simply not recognized anything out of place. But ironically, later in his search he suspected something remarkable that didn't require a star map, even though he failed to act on it. On September 29, before the news of Neptune's discovery had reached England, but just after the arrival of Le Verrier's prediction that the planet should display a notable disk, Challis had noted that one of the objects in their survey seemed too disklike, and this, again, turned out in retrospect to be Neptune. Yet with what seems to be a characteristic lack of urgency, he had deferred a closer inspection to a later night, and by that time the discovery of Neptune by German astronomers using French calculations was history.

Rev. W. T. Kingsley, who had been dining with Challis one evening at Trinity College in Cambridge in September 1846, later recalled an even more tantalizing story, which may have occurred the following evening. Describing his ongoing search over the table with Kingsley, Challis mentioned that one of the stars he had recently observed had appeared to show a disk. "Would it not be worth while to look at it with a higher power?" Kingsley suggested, and Challis agreed. Accordingly, after dinner, the two set out toward Challis's residence and the nearby observatory. Skies were clear, but before they could open the dome, Mrs. Challis offered to warm them up with a cup of tea, and by the time they had finished, clouds had rolled in, cutting short the evening's activities. "But for that cup of tea," according to Kingsley, "Adams would have had full credit." Kingsley claimed that their expedition to the observatory occurred in mid-September, but his memory was not precise, and Challis seemed not to have noticed the unusual appearance of Neptune until the end of the month. In any case, the event seems quite in keeping with Challis's laid-back attitude toward observing.

As battle lines developed in autumn of 1846, French pundits closed ranks around the understanding that Gallic science stood at the pinnacle of civilization and that Le Verrier's claim had to be defended against the encroachment of the British. In England, there was a feeling of populist indignation that the work of a brilliant young mathematician of humble origins had been ignored by the

English Scientific Establishment and scorned by the French. At the same time, there was a sense of public disappointment that a great opportunity for England had indeed been lost. A French magazine published a pair of cartoons that expressed the feeling on both sides. In the first, a British astronomer is depicted looking for a planet through his telescope, but he's clearly looking in the wrong direction to see a brilliant object behind him in the sky. In the second, the astronomer has indeed discovered the planet but only by pointing his telescope across the English Channel, where he reads an account of it in a French book. The French gloried in this version of the Neptune story; the English smarted in embarrassment.

Public attention in England focused, not surprisingly, on George Airy, who, as in his earlier dealings with Adams, seemed to have a talent for saying the wrong things at the wrong times. As the tempest over Neptune gathered force in England, Airy felt compelled to set the historical record straight as he saw it, sometimes at the expense of accuracy. On one day in mid-October 1846, he sent out separate letters to the three principal players in the Neptune saga, notable in their differences in style and substance.

In the letter to Le Verrier (which, along with several that followed, was widely published, to Airy's annoyance, in the French newspapers), Airy begins by lavishing praise on his French colleague, while trying to put in a modest claim for the English: "*You are to be recognized beyond doubt as the real predictor of the planet's place.* I may add that the English investigations, as I believe, were not quite so extensive as yours." But, in fact, as Airy acknowledged several years later, when he had actually read Adams' papers, the Englishman's analysis was every bit as sophisticated and thorough as Le Verrier's.

Writing to Challis, Airy was more supportive of England's claims: "Heartily do I wish that you had picked up the planet, I mean in the eyes of the public, *because in my eyes you have done so.* But these misses are sometimes nearly unavoidable." Sometimes, perhaps, but the English public was not so ready to dismiss this as an opportunity that did not knock loudly enough.

And finally, writing to John Couch Adams, Airy requested permission to publish all correspondence between them so that he could prepare a report on the Neptune affair to the Royal Astronomical Society. His reasons, Airy stated, were to set the record straight, to do justice to England, and to credit the individuals involved. Adams was only too eager to grant permission to Airy. Still, one gets the sense that Airy was well aware that he would come in for criticism, and that, if he published the correspondence with his own gloss, he'd be able to preempt any accusations of mistakes made on his part. Moreover, as a sign of his continuing ignorance of the man he was writing to, Airy addressed the letter to Rev. W. J. Adams, wrong on both the initials and the profession. (In his reply, Adams added in a postscript: "I may mention that I am not yet in Orders.")

A public inquiry of sorts took place in England on November 13, 1846, at a meeting of the Royal Astronomical Society. Airy presented his reading of the events leading up to the discovery of Neptune, Challis reported on his search program at Cambridge, and Adams described the methods he had used and the predictions he had made regarding the position of the new planet, taking care to note the chronology of his correspondence with Airy and Challis.

Airy's account is notable for its stiff formality and for its lavish praise of Le Verrier: "Since Copernicus . . . nothing (in my opinion) so bold, and so justifiably bold, has been uttered in astronomical prediction." Yet scarcely no praise is given to Adams, as if he were a peripheral player of no intellectual consequence. Airy still, it seems, did not realize the sophistication of Adams' work, and he was unwilling—probably for self-serving reasons—to grant that he should have acted more quickly to verify Adams' 1845 predictions.

Justifiably or not, Airy's story only inflamed ill-feeling toward him among the English public. Even Adam Sedgwick, a long-time friend of Airy and a highly respected geologist at Trinity College, had nothing but sobering words to offer in a letter shortly after the Royal Astronomical Society meeting:

> You were accused, not only of unreasonable incredulity and apathy towards Adams of St. John's, but also of having (as was said) "snubbed him from the first"

and so acting on a timid person prevented him from reaping the honours of great discovery. . . . [H]ad the results communicated to you and Challis been sent to Berlin . . . the new planet would have been made out in a very few weeks, perhaps a very few days, and the whole business settled in 1845—Adams the sole, unadvised, unassisted discoverer.

Airy answered with self-justifications and excuses but admitted that he would have to "wipe my bloody nose quietly at home." In his autobiography, published posthumously, Airy devotes only one line to the entire Neptune affair: "I was abused most savagely both by English and French." He refers readers who want more details to his 1846 report to the Royal Society; the incident was clearly too painful to go over again.

Challis had an even more difficult time scuffling out from under the cloud of suspicion that he was largely responsible for not finding Neptune sooner. After all, he had been Adams' confidant and first line of contact, and he had possessed the position of the planet for almost a year without acting on it. Reviewing Challis's report, a member of the Royal Astronomical Society commented that it was a "pitiful story." He added, "To do it justice, it is candid. No one would dream of doubting its veracity, for what could induce any man to produce a tale of that complexion."

The only one who came out of the November meeting smelling good was Adams, whose modesty and candor had made an impression that was to set him on course to an honored and respected place in the scientific community. But damage had been done. At the end of November 1846, the Royal Society of London awarded its Copley Medal—still regarded as one of the highest honors in science—to Urbain Le Verrier in recognition of his discovery of Neptune. The citation bore no a mention of Adams. Le Verrier was also made a fellow of several scientific societies in Europe, and the King of France named him to the Légion d'honneur. He became, in time, the most highly influential mathematical astronomer of the century, a gray eminence among European astronomers.

John Couch Adams, on the other hand, only received recognition slowly. The Royal Society awarded him the Copley Medal two years after Le Verrier, and he was eventually rewarded with several academic positions, eventually replacing Challis as director of the Cambridge Observatory. On a visit to Cambridge in 1847, Queen Victoria offered him a knighthood, which he declined, claiming that he did not wish to be compared with Newton, who had received a similar honor. And ironically, on Airy's retirement in 1881, Adams was offered the post of Astronomer Royal, which, perhaps with a wry smile at the strange convolutions of fate, he also turned down. Throughout his life, however, he refused to dwell on any presumed injustice, and he spoke only with modesty about his part in the discovery of Neptune.

None of the controversy that so poisoned public discourse about Neptune right after its discovery seemed to have any effect on personal relations between Adams and Le Verrier. In June 1847, when Le Verrier attended a meeting of the British Association for the Advancement of Science, the two had a chance to meet face-to-face, and according to one witness, "I remember . . . being charmed, like everyone else, at the cordial handclasp that was exchanged." The two remained warm friends and correspondents for the rest of their lives. (Le Verrier died in 1877 and Adams in 1892.)

Over the years, public controversy over the discovery of Neptune faded, but the French/British rivalry for credit in the find occasionally resurfaces when Neptune is in the news. (Still, the Neptune flap is much less acrimonious than many other national rivalries; the invention of the telegraph, for instance, is still hotly contested among many countries.) For several decades, however, Johann Galle was accorded major credit for first spotting the planet. Only after the death of Heinrich d'Arrest in 1877 did Galle make clear the young man's crucial role in the discovery—it was d'Arrest who suggested using the new Bremiker map, and it was d'Arrest who sat next to Galle, meticulously cross-checking the stars to see if any were out of place. d'Arrest, who later became a noted astronomer, was only a graduate student at the time of his work in Berlin and perhaps expected no more credit than he got. In any case, the stakes were smaller, since it was generally accepted that, without the predicted position to go on, Neptune would not have been discovered as quickly as it was.

NEPTUNE, AN ABIDING MYSTERY

Neptune eventually took its place among the pantheon of major planets, with new generations of astronomers unaware or unconcerned about its stormy discovery. Still, there were nagging questions about the scientific significance of the newfound planet. No one doubted that the work of Adams and Le Verrier represented a great achievement of Newtonian theory, but there were troublesome discrepancies between the planet that had been hypothesized and the planet that actually was discovered. Both Adams and Le Verrier had predicted a rather massive planet located about 36 astronomical units from the Sun in a slightly elliptical orbit, but when enough observations had been gathered to determine its characteristics, Neptune was found to be about half as massive as theorized, and its orbit was nearly a perfect circle with a radius of only 30 astronomical units. (This, interestingly, was a gross violation of Bode's law, which would have predicted a value of 38.8 astronomical units; it was the first time that Bode's law had been so grossly "violated," and it confirmed some astronomers' suspicions that the law itself was simply a numerical accident.)

Did the discrepancy between the theoretical planet and Neptune indicate that something was wrong with the computations? It was possible that the new planet had been discovered purely by accident, although the fact that both predictions agreed so well with each other and with the actual position of the planet made it seem likely that Newton's laws had played a crucial role in at least pointing astronomers in the right direction. It seemed more likely, given the experience of the past, that Neptune was not behaving properly because it (and Uranus) were still being acted upon by forces that had not fully been accounted for. Perhaps there was yet another planet beyond Neptune that was tugging it in unanticipated directions. If so, mathematical astronomers felt that they now had the tools to discover it, but it was to take nearly another century before they had a chance to put their calculations to the test. The search for that next planet would ignite a new controversy that would reach its climax in the angry discussions at Prague in August 2006.

CHAPTER 6

Vulcan, the Planet That Wasn't

As I was going up the stair
I saw a man who wasn't there
He wasn't there again today
Oh, how I wish he'd go away

—Hughes Mearns, "Antigonish," 1899

The failure to find a planet can be just as contentious as a successful discovery. In fact, some of the liveliest astronomical arguments have centered on celestial objects that never existed at all. The most remarkable of these phantoms bore a name everyone knows: Vulcan, the home of Mr. Spock of *Star Trek*. Spock's planet, of course, is science fiction, a world supposedly orbiting the star 40 Eridani, 16 light-years from Earth. The Vulcan of astronomical history, however, was a planet that many astronomers in the latter half of the 1800s believed orbited our Sun, somewhere inside the orbit of Mercury.

For more than a decade, astronomers mounted serious searches for the planet at every favorable opportunity. Following several claims that it had been discovered, and a multitude of failures to spot it at all, the Vulcan of the 1800s turned out to be as fictional as the one of *Star Trek*.

MERCURY'S PUZZLING PERIHELION

Despite impressive successes in predicting the motion of Halley's comet in 1758 and in sniffing out the presence of Neptune in 1846, lingering doubts about the universal effectiveness of Newton's theory of gravitation still haunted the astronomical community. Over two centuries since Newton, the greatest mathematicians in Europe had developed techniques to account for the gravitational attraction of all the known objects in the solar system in calculating orbits. Yet notable disagreements between theory and observation remained, the most troublesome of which was a small but mysterious deviation in the orbit of Mercury.

The closest planet to the Sun, Mercury has the most elliptical orbit of all the planets except Pluto. Though its average distance from the Sun is about 36 million miles, it swings out as far as 43 million miles to the *aphelion* of its orbit and then loops in at closest approach, or *perihelion*, to a mere 28.5 million miles. If Mercury and the Sun were the only bodies in the solar system, the planet would retrace this same oval path over and over again. The combined gravitational attraction of the other planets and the Sun, however, causes the axis of Mercury's orbit to change direction, shifting the position of perihelion each time it goes around. If you traced the orbit on a piece of paper it would resemble the petals on a daisy. The shift in the perihelion is small, barely a thousandth of a degree each year, and it takes over two and a half million years for the orbit to return to the same orientation, completing all the petals on the daisy. Yet, small as it is, by careful observations of the position of Mercury, astronomers can accurately measure the shift, which is called the *perihelion advance*, or *precession of the perihelion*, without waiting a couple of million years.

There was no reason to believe that gravitational theory would fail to predict the exact shape of Mercury's orbit, and it was one of the first problems that young Urbain Le Verrier tackled when he shifted his profession from chemistry to astronomy. In 1843, three years prior to his discovery of Neptune, he published the results of his analysis of Mercury. The positions he calculated were close to the measured orbit, but there was an unexplained difference between the observed and calculated rate of perihelion advance that was too large to be shrugged off as within the bounds of measurement error. The calculations gave a value that was too small by about a hundredth of a degree per century. Something was wrong, but at the time, Le Verrier was inclined to attribute it to an error in the available data on the masses of the planets or a slip in logic or arithmetic on his part—he was, for all his talent and youthful confidence, relatively new at the game. When the matter of Uranus's orbit captured his attention in 1845, he dropped the study of Mercury for a while. Then, immediately following the discovery of Neptune, he had to deal with public adulation and the controversy over priority, as described in Chapter 5.

LE VERRIER'S SOLUTION

Fame brought with it added responsibilities that further delayed Le Verrier's work on Mercury. In 1854, he took over as director of the Paris Observatory and occupied himself for several years with a drastic reorientation of the activities of the observatory's mission. His predecessor, François Arago, who had died the year before, had been a colorful and imaginative character who ran what Le Verrier must have regarded as a rather loose ship. Scientifically, Arago was noted for his work in both astronomy and optics—it was he who had put Le Verrier on the track of Neptune by suggesting an analysis of the orbit of Uranus. The public knew Arago as a writer of engaging popular science books and as a radical who had manned the barricades during the Paris Commune uprising of 1848. Le Verrier, in contrast, was a political conservative, rather formal in his writing style, and, for all his mathematical genius, a narrow thinker who focused single-mindedly on analyzing planetary motions. Astronomy to him was not a grand narrative about the universe but an actuarial challenge. He set the Paris Observatory staff to making systematic measurements of planetary positions and to carrying out the laborious computations required to match theory with observation. There was to be no nonsense under his regime; there was important work to be done.

Mercury's odd behavior was always in the back of his mind, for he saw himself no longer as a neophyte astronomer but as a master of celestial mechanics, the man who would tidy up all the loose ends of Newtonian theory. And so, five years into his directorship of the Paris Observatory, Le Verrier took the time to sit down and look carefully at Mercury once again. He was older and more accomplished and had a longer baseline of observations to compare, but his results were still troublesome.

Discrepancies between theory and reality were particularly noticeable in observations of Mercury's transits, rare occurrences when the planet was seen crossing the face of the Sun. The starting times of these events (called "first contact," when the dark silhouette of Mercury first makes its appearance on face of the Sun) could be measured with exquisite accuracy, leaving no doubt that any disagreement between theory and observation was a result of a failure of theory, not carelessness on the part of the observer. Understandably, predictions of the times of first contact by the first Newtonian astronomers had been way off—Edmond Halley's table for a transit in 1753 was almost an hour in error. But mathematical techniques and a general knowledge of masses and distances in the solar system had improved over the years. By the Mercury transit of 1843, astronomers only missed the time of first contact by 16 seconds. Still, Le Verrier could not be satisfied with a 16-second error, and as he embarked on his recalculation of its orbit 16 years later, he believed his analysis would put him right on the money. He was to be disappointed: try as he might, he could not reduce the discrepancy significantly below 16 seconds.

He was not discouraged. The discovery of Neptune left him confident that Newton's laws were correct and that the only explanation for the persistent error was—as had been the case in 1846—the presence of a previously unknown member of the solar system. In September 1859, he sent a letter to the French Academy of Sciences describing his latest work on Mercury and his conclusion that new worlds remained to be discovered in the gap between Mercury and the Sun. The unseen mass might be a single planet, or it might be a swarm of smaller bodies, like the belt of asteroids between Mars and Jupiter.

Why had this supposed intra-mercurial mass so far escaped the attention of astronomers? Simply—so Le Verrier was convinced—because of the difficulty of discovering anything close to the Sun. Mercury itself is the most troublesome of planets to observe; because of its Sun-hugging orbit, it is only visible low in the sky for an hour or so before dawn or after sunset. A small planet or asteroid closer to the Sun than Mercury might only appear in a dark sky for a few fleeting minutes and might be obscured by clouds, trees, or mountains. It was not difficult to imagine that such an object could go completely unnoticed by generations of sky watchers.

Le Verrier was unable to give astronomers detailed instructions on where to point their telescopes; he didn't have enough information to calculate a position as he had for Neptune. But he sensed that he was on track to his second great cosmic discovery. In his letter to the Academy, he urged astronomers to organize systematic searches for intra-mercurial objects. The most effective route to success, Le Verrier noted, was to look for objects close to the Sun during total solar eclipses or to observe the intra-mercurial planets in transit across the face of the Sun.

Le Verrier, to be sure, was not the first person to suggest that there might be planets closer to the Sun than Mercury. Astronomers had been watching the Sun's surface through telescopes ever since the early 1600s, and though most of the dark markings they saw on its face were sunspots—which rotate with the Sun once every 25 days or so—there were dozens of random reports of fast-moving dots crossing as well. None had been definitely identified with a planet, but astronomers were alert to the possibility that one or more of these sightings was real and that a systematic search for an innermost planet might be worthwhile. An amateur astrono-mer, Heinrich Schwabe, began such a search in 1826 and faithfully charted the Sun virtually every day for three decades—no mean feat, considering the prevailing weather at his site in Dessau in central Germany. Schwabe saw no planets transiting the Sun during the long span of his solar observations, but he could plainly see what other less-dedicated researchers had missed: that the number of sunspots rose and fell once every 11 years or so. He had discovered the sunspot cycle. Schwabe was elated: "I may compare myself to Saul, who went out to seek his father's asses, and found a kingdom." But given the constant scrutiny he had devoted to his observations of the Sun, the prospects for the existence of intra-mercurial planets seemed dim.

Despite the failures of previous surveys like Schwabe's to turn up any systematic evidence of planets closer to the Sun than Mercury, Le Verrier's call to action could not be ignored. He was the high priest of Newtonian physics, after all, the man who had expanded the boundary of the solar system by sheer force of logic. If Monsieur Le Verrier thought there was an unknown body lurking within Mercury's orbit, then whoever failed to join the search risked missing out on the glory of discovering the next big thing in astronomy. And then there was the curious matter of Dr. Lescarbault.

VULCAN ENTERS THE SCENE

Late in December 1859, Le Verrier received a letter from a country physician, Edmonde Modeste Lescarbault, from the small town of Orgères-en-beauce, about 50 miles southwest of Paris. An eager amateur astronomer, Lescarbault split his time between patching up patients and scanning the sky with a modest-sized refracting telescope. Since the early 1850s, he'd been looking at the solar disk, like Schwabe (but more sporadically), with the intention of catching sight of a transiting planet if one existed. On March 26, 1859, by his account, he was surprised to see a speck moving across the face of the Sun, far too rapidly to be a sunspot. He'd filed it away as a curiosity, not certain whether that one observation constituted sufficient evidence to claim discoverer's rights, until, half a year later, he read a report of Le Verrier's analysis of Mercury and his conclusion that there was a planet or planets closer to the Sun. Could that have been the dark shadow he had seen in March?

Le Verrier had never heard of Lescarbault, needless to say, and the director of the Paris Observatory was not about waste time on every crackpot letter he received. It was odd that Lescarbault had not announced the March sighting earlier, and it seemed likely that the letter writer was simply a crank who wanted to attract the attention of a noted public figure. But Le Verrier could not resist the possibility that the predicted object really existed. So on December 30, he paid a visit to Orgères-en-beauce and knocked unannounced at Dr. Lescarbault's door, fully expecting to meet an unabashed charlatan whom he could dismiss with a stern lecture. He was surprised by the timid and accommodating man who answered. The country doctor was hospitable and, though intimidated by the presence of the great man, was able to respond satisfactorily to Le Verrier's grilling. The astronomer returned to Paris convinced that Lescarbault's observations, though far from precise, were trustworthy enough. At a meeting of the Paris Academy on January 2, Le Verrier gave a report on his visit and on Lescarbault's results, declaring an "absolute conviction that the details of his observations ought to be admitted to science."

Neptune was still a fresh memory, and the press was quick to crow over this second apparent triumph—this time a clean sweep by the French. Le Verrier made the rounds at Paris parties, and Emperor Napoleon III (at Le Verrier's suggestion) awarded the French Legion of Honor to Lescarbault a few weeks later. An article in the *North British Review* at the time described public sentiment: "Garibaldi and the weather ceased to interest the Parisians; and the village doctor, in his extempore observatory, and his round black spot . . . were the only subjects of discussion." The Royal Society of London was lavish in its praise, "applauding this second triumphant conclusion to the theoretical inquiries of M. Leverrier." However, Lescarbault's discovery, as it turned out, was far from a conclusion to the problem of Mercury's perihelion advance.

Le Verrier called the presumed new planet Vulcan—a name that the French astronomer Jacques Babinet had coined in 1846 for a hypothetical body he thought might exist inside the orbit of Mercury. Vulcan, the Roman god of the forge and of volcanoes, was an appropriate name for a body so close to the Sun. From the meager information Lescarbault had provided, Le Verrier was able to estimate that Vulcan's orbital period was around 19.7 days. Its precise distance from the Sun was uncertain, but its mass came out to be about 15 percent of the mass of Mercury. Such a small object could not account for all the perturbations that afflicted Mercury's orbit, he realized; to produce the observed perihelion advance required about 20 objects this size. Still, it was possible that Lescarbault's planet was just the largest of what would turn out to be a belt of planetoids in similar orbits. As far as finding intra-mercurial objects, the more the merrier. Things were looking good for spotting Vulcan and its cohorts now that there was some idea on where to look.

It still would not be easy. Though Le Verrier estimated Vulcan would appear as a moderately bright star, anything that close to the Sun was apt to be lost in the glare of daylight. The best times to see the planet would be when it was crossing the plane of the Earth's orbit, because then, if the Earth was in the right place, the dark silhouette of the planet would be evident against the face of the Sun (though one would have to be careful not to be misled by sunspots). Le Verrier calculated that the propitious seasons for Vulcan hunting would be around the beginning of April and the beginning of October. The astronomical community was quick to respond. At the next favorable opportunity, between March 29 and April 7, 1860, astronomers worldwide were watching, examining the solar disk for the slightest sign of a blemish.

They saw nothing, as was to be the case during other Vulcan searches over the next two decades (with a few notable exceptions, as we shall see). The negative result was disappointing to Le Verrier, but it came as no surprise to Emmanuel Liais, a Frenchman who was at the time working for the government of Brazil. Described by a contemporary as an astronomer "of considerable skill," Liais had been observing the Sun at the very same day and hour as Dr.

Lescarbault and could attest to the fact that no unusual spot, speck, or blemish had appeared crossing the solar disk. In a report published in the prestigious German journal *Astronomische Nachrichten* just after the 1860 observing campaign, he emphatically stated his conclusion: "I am in a position to deny, and to deny positively and absolutely, the passage of a planet across the Sun at the time indicated." A few years later, he devoted an entire chapter of his popular book, *L'Espace Céleste*, to "The Absence of Planets Close to the Sun."

VULCAN IGNITES AN ASTRONOMICAL BONFIRE

In both his technical article and popular writings, Liais challenged the very notion of a planet or planets inside the orbit of Mercury, drawing the battle lines of a controversy that was to simmer—and occasionally boil over—in the decades to follow. He argued that if Vulcan existed, it should surely have been seen as a starlike object close to the Sun at solar eclipses. Granted, eclipses are rare and their duration short, but if astronomers missed seeing it at every eclipse, there was no way the planet could have been missed at the other favorable situations for discovery: in transit across the face of the Sun. Given the short period of its orbit—only a few weeks by most estimates—these crossings would be frequent, and since transits last several hours and are visible over the entire daytime hemisphere of the Earth, it is inconceivable that astronomers would not have seen many transits of planets inside Mercury. Thus the absence of evidence for Vulcan was, to Liais, evidence of absence.

Liais cautioned that Le Verrier's trust in the precision of astronomical measurements was misplaced. It was premature to "deduce" a planet from the discrepancy between theory and observation, because there might well be hidden errors in the observations, not to mention errors in calculation. (This all-too-human trait persists to this day, no doubt. In his 1969 memoirs, American astronomer Harlow Shapley quipped, "A hypothesis or theory is clear, decisive, and positive, but it is believed by no one but the man who created it. Experimental findings [observations], on the other hand, are messy, inexact things, which are believed by everyone except the man who did that work.")

Finally, Liais questioned the veracity of Lescarbault himself, a claim that, in retrospect, was unsubstantiated. It is far more likely that the doctor was mistaken or misled, as Liais himself suggested, by some other phenomenon, such as a high cloud passing over the face of the Sun. But there were doubtless later "sightings" of Vulcan that were fabricated by cranks or people seeking attention, often sparked by articles about it in newspapers and popular magazines. Englishman Richard Proctor, a widely read popular astronomy writer, reported receiving a number of letters from people claiming to have seen the planet. The crank letters were usually recognizable by descriptions of objects far too big, bright, or fast-moving to be believable. In one case, a humbug astronomer from Louisville, Kentucky, to substantiate his pet theory on how the planets affected the weather, claimed to have seen Vulcan at a time when it would have been on the opposite side of the Sun (if it existed). "To see Vulcan through the sun," Proctor wrote, is "a very remarkable achievement, indeed."

The attack on Vulcan by Monsieur Liais won the support of a number of noted scientists, among them French astronomer Camille Flammarion and an American, Christian H. F. Peters. Vulcan was becoming a hot-button issue, and a chorus of harsh criticism rumbled through the astronomical community. But the more Le Verrier was assailed, the stronger was his resolve. He reaffirmed his trust in Lescarbault's discovery and began to see his critics as personal enemies. Still, had there been no further evidence for Vulcan, the factions might have simply tired of arguing, and the hypothetical planet might have slowly faded from the scientific agenda of the time. As it was, occasional reports from observers fueled the debate for another 20 years or so.

In March 1862, for instance, amateur astronomer M. W. Lummis in England claimed to have seen Vulcan in transit. Despite two other reports that the object in question was a sunspot, Le Verrier and his supporters used Lummis's data to reestimate Vulcan's orbit in anticipation of future sightings. In 1869, American astronomers looked for the planet at a total eclipse of the Sun visible from Iowa, and though all the professional astronomers reported looking carefully and seeing nothing, there was one rather unspecific account of a "minute object seen near the sun" by four naked-eye observers in Sioux City. The encouragement that this casual sighting might have given to the pro-Vulcan camp was tempered by the negative reports of numerous English amateurs who participated in organized campaigns to see Vulcan in transit across the Sun in spring 1869 and 1870. Though many sunspots were charted, there was no trace of a planet.

Faith in Vulcan had reached a low ebb by 1876, when on April 4, Heinrich Weber, a German astronomer working in China, sent a telegram to colleagues in Europe that he had just observed a round object in transit across the Sun. It was just the right time of the year for an appearance of Lescarbault's planet, and it sent astronomers reflexively scurrying to their telescopes. By the time the telegram arrived, however, the planet seemed to have disappeared.

Le Verrier, by then in his mid-60s, was elated by the news and set to work redoing his calculations. The game was afoot again! Based on the new data and the new computations, he revised the period of the planet to 33 days and put out the call for a transit patrol in October. Again, telescopes around the world turned toward the Sun, and again the planet was not seen. Undaunted, Le Verrier made more predictions for transits in 1877, ignoring the discouraging news from Spain that Weber's supposed planet was really a nearly circular sunspot. Astronomer Royal George Airy later sent photographs from Greenwich confirming the Spanish report, but Le Verrier persisted. Once again, in March 1877, though they looked, astronomers saw nothing crossing the face of the Sun. Le Verrier, who was in failing health at the time, could not have been happy. When he died in September 1877, his eulogies described the great man as an astronomer of the highest order and as the discoverer of Neptune. Colleagues glossed over his frustrating campaign to calculate the perihelion advance of Mercury and his insistence that there were planets inside Mercury's orbit. Better to dwell on his accomplishments than his less-than-magnificent obsessions.

VULCAN'S LAST HURRAH

The last chapters in the Vulcan debate had not yet been written. Matters came to a head on July 29, 1878, when a total eclipse of the Sun swept across the western United States. For months before, American astronomers had been planning expeditions to points along the track of totality in Wyoming, Colorado, and Texas. They were eager to see if recently developed photographic techniques could be used to record the event effectively, and they recognized a unique opportunity for U.S. science—the possibility of finding Vulcan from American soil. Not yet recognized as a scientific power, America would nevertheless succeed where Europe had failed, and newspapers were abuzz with feature articles on the upcoming search for Vulcan. The Government Printing Office issued detailed instructions for observers of the eclipse, including a three-page foldout map of stars in the vicinity of the Sun on the day of the eclipse so that anyone spotting a suspect dot of light during totality could immediately identify it as a planet.

The weather at most spots in the western United States was ideal for viewing on eclipse day, and observers made the most of their few minutes in the shadow of the Moon to carry out experiments as planned. Simon Newcomb of the U.S. Naval Observatory was near the rail line at Separation, Wyoming. Until three minutes before the total phase of the eclipse, he remained inside a darkened room so that his eyes could become accustomed to the dim light of totality (a tactic he had also employed in Iowa in 1869). Then, walking to the telescope, he took a short look at the pearly glow of the corona, the outer atmosphere of the Sun, and began to scan the sky around the Sun for unfamiliar objects. As had been the case in 1869, he saw absolutely nothing of interest. Neither did his colleagues E. J. Holden in Central City, Colorado, and David Todd in Dallas, Texas. For the most part, all the observers at the 1878 eclipse who took the trouble to look for planets submitted routine negative reports. They saw plenty of stars but none that weren't already on the charts.

Yet just a few feet away from Simon Newcomb at the Separation, Wyoming, station was James C. Watson, a brilliant young astronomer from the University of Michigan, Ann Arbor, who saw what all the others had somehow missed. It was a rather bright star that was not on the charts—a possible planet! He carefully noted its position "in reference to the sun and a neighboring star . . . by a method which obviates the possibility of error." Then he ran over to Newcomb, whose telescope was nearby, and quickly blurted out the news. Totality was about to end, and Watson wanted Newcomb to verify the position of the object, but Newcomb had one last measurement to make and could not comply immediately. By the time he finished, the sky was light again. "It is of course now a matter of great regret," wrote Newcomb, "that I did not let my own object go and point on Professor Watson's."

Regret indeed. This brief sighting by Watson, this tiny point of light, reanimated the debate over Vulcan as nothing before. Lescarbault had been an amateur, as had most previous observers of suspect objects near the Sun. But Watson was one of American's most promising young astronomers, respected for both his mathematical skills and his sharp eye at the telescope. He had discovered 20 asteroids, aided by an uncanny photographic memory for star charts that enabled him to recognize immediately any object that seemed out of place. Even without Newcomb's corroboration, Watson's claim to have found Vulcan could not be brushed aside as a fabrication or hallucination. Watson was, granted, an enthusiastic believer in the reality of Vulcan, but all his previous work had been of the highest integrity. His observation, moreover, was given added weight by another positive sighting at the eclipse from Denver by Lewis Swift, an amateur astronomer from Rochester, New York, who was known as the discoverer of several comets. Swift reported seeing two uncharted stars, actually, both about three degrees southwest of the Sun. The objects seemed to be roughly in the same place as Watson's.

News of the new planet exploded in the press. The *Laramie Daily Sentinel* announced that "Professor Watson, of Ann Arbor, Michigan, who is now the most noted astronomical observer and discoverer in the world, had taken the job of FINDING VULCAN. And he found it." An article that ran in the *New York Times* of August 16 exulted,

"The planet Vulcan, after so long eluding the hunters, showing them from time to time only uncertain tracks and signs, appears at last to have been fairly run down and captured." If this was Vulcan, however, why had so few people seen it, especially since so many were looking for it? Presumably because they had not been looking at the right place. The eclipse was only a few minutes in duration, after all. Anyone who has experienced a total eclipse of the Sun can attest to how quickly time seems to fly by during that short period, and observers are prone to making errors in their haste. Watson, moreover, was legendary for being able to recognize patterns in the heavens and for seeing stars fainter than others believed possible. Most likely a singular combination of luck and visual acuity had led him to the discovery.

Yet after the first rush of acclaim, Watson's triumph began to create a substantial backlash among astronomers who did not share his enthusiasm for Vulcan. Doubts crystallized after the 1879 publication of a critical article by astronomer Christian H. F. Peters from Hamilton College in central New York. Peters had pooh-poohed the existence of Vulcan from the very beginning, questioning the reality of Lescarbault's observations and Le Verrier's predictions. He saw the 1878 eclipse reports as an unequivocal vindication of his skepticism, "because the searches were made very thoroughly, and among the ablest astronomers of the country . . . only two observers report to have seen something." Of those two observations, Peters remarked, only Watson's was worthy of scientific consideration. Swift did not report his sighting until after hearing of Watson's, thereby tainting his own report with suspicions of bias. And besides, the two astronomers' estimates of the alleged planet's position were not really consistent with each other. Watson's measurements, furthermore, were of questionable accuracy, since Watson could only view his measuring scale in the dim light of totality and had likely overestimated how reliably he could read the dial setting on the telescope. According to Peters, what Watson thought was an unknown planet was actually a well-known star, Theta Cancri.

Peters did not level his cannons only at Watson and Swift. All previous reports of supposed planets in transit were also specious, he argued. Sunspots were easy to confuse with planets, and no experienced sunspot observer had ever reported a suspicious object. "With few exceptions, only persons otherwise wholly unknown as astronomical workers have been favored with a view of the mythical planet," he wrote. "Most of them even cannot be called observations, in an astronomical sense, consisting only of rude general statements, without precision, often contradictory in themselves, always incomplete."

Watson and Swift defended themselves vigorously in the professional journals. They were experienced observers, they protested, not prone to making simple errors of measurement, and they certainly would not have mistaken the star Theta Cancri for a planet. They were convinced of the validity of their claims, and they took offense at Peters' implication that they had doctored data to make themselves—and Vulcan—look good. But they could not undo the damage; Peters' criticisms, though harsh and needlessly ad hominem, went to the heart of the scientific matter and, in the minds of all but a dedicated few, served to undermine the already shaky foundations of the entire Vulcan enterprise. From this time on, Vulcan's orbit was—figuratively speaking—an inexorable death spiral into the dustbin of history.

The supposed planet took a few final bows in the next few decades, but its erratic behavior seemed to confirm its mythological status. In January 1880, there was a lone sighting at a total eclipse in California, but nothing was seen at an eclipse in Egypt in 1882 or at an eclipse in 1883 on Caroline Island in the Western Pacific. As photography came into wider use, astronomers no longer had to rely on visual scans made in the haste of totality. Photographs taken at a 1904 eclipse in Spain and in the South Pacific in 1909 showed no trace of planets down to magnitudes much fainter than the objects Watson and Swift reported. Astronomers confidently concluded that there was no large or even moderate-sized planet located inside the orbit of Mercury, and searches for such objects were gradually eliminated from the research agendas at solar eclipses. It was possible, of course, that there were some objects, too small to show up on photographs, that were in orbit around the Sun. But even so, as W.

W. Campbell wrote in an article entitled "The Closing of a Famous Astronomical Problem" that appeared in *Popular Science Monthly* in 1909, "any such bodies would fail hopelessly to supply the great mass of material demanded by Le Verrier's theory."

THE PERIHELION PROBLEM SOLVED

So what was wrong with Mercury's orbit? What was causing the tiny but undeniable discrepancy in the precession of its perihelion? The answer, in the end, turned out to be the alternative that Le Verrier rejected as virtually unthinkable: Newton's theory of gravity was flawed. Albert Einstein solved the problem almost as an afterthought to his general theory of relativity, an elegant mathematical system that radically altered commonly held notions of time and space and that reformulated the law of gravitation in a form that Newton would not have recognized. The orbits of planets were no longer viewed as masses accelerating around the Sun under the influence of a gravitational force. Instead, space-time was seen as a geometrical reality, the fabric of which was distorted by presence of matter. Planets orbited in the distorted space-time around the Sun following curves called geodesics, the shape of which could be computed using elegant tools of a branch of mathematics called variational calculus.

Einstein developed general relativity as a solution to what he saw as philosophical inconsistencies in Newton's physics, not as a way to better describe the orbits of planets. He was trying to find a way to describe how different observers measure space, time, mass, and other physical properties of objects when gravitational fields or accelerations are present. General relativity, to Einstein, solved a conceptual problem—the question of "general covariance" in his terminology. But when the orbit of Mercury was calculated using general relativity, its perihelion advance was in perfect agreement with the observations. No intra-mercurial planets were needed.

In 1915, in a letter to a friend about his recently completed work on general relativity, Einstein wrote: "The explanation of the shift in Mercury's perihelion, which is empirically confirmed beyond a doubt, causes me great joy, but no less than the fact that the general covariance of the law of gravitation has after all been carried to a successful conclusion."

VULCAN'S REALM TODAY

With the publication of Einstein's theory, Vulcan lost its raison d'être. But astronomers did not totally abandon the search for objects closer to the Sun than Mercury. Although we no longer need a large amount of matter to perturb the orbit of Mercury, it's always possible that small objects very near the Sun could so far have escaped detection.

We already know of some. The Solar and Heliospheric Observatory (SOHO), launched in 1995, is a spacecraft that orbits between the Earth and the Sun. Equipped with a battery of telescopes that continuously monitor the Sun and the region around it, it is ideally situated to detect small bodies near the Sun that can't be seen from the Earth's surface. As of 2009, SOHO had detected 1,500 small comets close to the Sun, most of them about to crash into the Sun and be destroyed. Comets are loose aggregations of ice and gravel, and as they approach the Sun, they often come apart, leaving behind a trail of debris. These sun-grazing comets, of course, don't have any significant effect on planetary orbits, and they are evanescent phenomena. Though they are so small that it is doubtful they could have been seen in transit across the Sun, it's possible that one or more of these might have been responsible for some of the supposed sightings of Vulcan at eclipses in the 1800s.

As to anything more permanent than kamikaze comets, the jury is still out, but there are reasons to believe that if there's any material orbiting inside Mercury's orbit, it is very small and very sparsely distributed. Computer simulations by astronomer Alan Stern and collaborators at the Southwest Research Institute in Boulder, Colorado, have revealed that a belt of asteroids near the Sun would disappear relatively quickly, some objects colliding and falling inward and other debris being dispersed into space by the intense radiation from the Sun. No more than a few hundred objects larger than one kilometer could have survived, amounting to a total mass a tiny fraction of that which Le Verrier assigned to Vulcan.

The supposed planet has truly disappeared, surviving only in the name astronomers assign to the smattering of tiny objects—yet to be discovered—that might possibly populate the innermost reaches of the solar system. Astronomers call them "vulcanoids." In 2002, Stern and his colleague Daniel Durda conducted a search for vulcanoids using an ultrasensitive electronic camera designed to record objects 600 times fainter than can be seen with the naked eye. To eliminate the blue glare of the Earth's atmosphere, they hitched a ride on a high-flying aircraft, an F/A-18B jet fighter flying at close to the speed of sound. As twilight gathered in the west, the astronomers and their pilot sped through the blackness of near-space and recorded over 100,000 images of regions close to the Sun, which they later examined using computers, looking for faint orbiting objects. It was technology undreamed-of by those who had searched for Le Verrier's Vulcan in the 1800s, but the result was the same. They found nothing.

CHAPTER 7

The Pluto Saga Begins

What a thrill it was to me, and I realized we were looking
farther out in the solar system than anyone had ever looked
before. It was also disappointing—the planet was a very faint,
unimportant looking 15th magnitude star-like point.

> —Clyde Tombaugh, "The Discovery of Pluto: Some
> Generally Unknown Aspects of the Story," 1986

The discovery of Neptune by an Englishman and a Frenchman in
1846 opened a new era in planetary discovery and, ironically, a
new era in planetary controversy. To be sure, Neptune's position
had been predicted, not by the sketchy numerology of the Titius-
Bode relation, but by a combination of bedrock Newtonian physics
and meticulous observation. Pushing outward beyond Neptune,
however, would lead astronomers into a labyrinth of mathematical and observational difficulties. What is more, the lust for planet
finding attracted of one of the most controversial figures in the
history of astronomy, a wealthy Bostonian named Percival Lowell,

who combined the mathematical rigor of a professional astronomer with the obsessive zealotry of an amateur and who had virtually limitless financial resources to devote to his passions. His efforts to find a new planet where others had failed led to the discovery of Pluto in 1930 by a country lad from Kansas named Clyde Tombaugh. And we are still debating today what that discovery signified.

Still, in the decades immediately following the work of Le Verrier and Adams, the discovery of farther planets at the edge of the solar system seemed preordained. Since a clever application of Newton's law of gravitation had predicted a previously unknown member of the solar system, the same method ought to be able to detect other unseen planets farther out. Yet though most astronomers in the mid-1800s took it for granted that continued observations would reveal planets beyond Neptune, they were in no great hurry to go out and find them.

Discovering the next planet, according to conventional wisdom, would take decades, and it was futile to engage in a search until enough time had passed so that the accumulated evidence clearly defined the planet's position. To emulate Adams and Le Verrier, astronomers first had to make precise theoretical calculations of the path of the outermost known planet, Neptune in this case, taking into account the gravitational force of all the inner planets. When calculated positions were compared with observations, discrepancies might appear that would point in the direction of a previously unknown object orbiting beyond Neptune. For instance, if Neptune seemed to be orbiting a bit farther out than predicted during part of its orbit, this would indicate that an unaccounted-for planet was pulling it in that direction.

It was easier said than done. Theoretical predictions, in the first place, could not be made with the push of a button. In the days before electronic computers, lengthy calculations, made by hand, were tedious and fraught with error. A single mistake in a long chain of manipulations could lead to an accumulation of incorrect results. Researchers might work for months, writing down sums and products and quotients only to produce a table of numbers whose reliability was uncertain at best. Patience and careful cross-checking were their most important tools.

But even had they been able to produce error-free calculations, astronomers also had to observe a planet for a substantial fraction of its orbit before any presumed differences became severe enough to be noticed. Because Uranus orbits the Sun in 84 years, Neptune was not discovered until more than a half-century of Uranus's orbit had been traced out. Small deviations in its orbit were suspected shortly after Uranus was discovered, but they did not become really evident until the 1830s. In addition, though astronomers did not know it until after Neptune was discovered, Uranus and Neptune were closest to each other in the 1820s, and so the deviations of Uranus were greater just after this time than they had been at any time since its discovery.

The situation got more difficult the farther out one went in the solar system. Neptune has almost twice the period of Uranus—almost 165 years—so astronomers estimated that decades of observations were needed before its orbit could be tested against theory. Even master mathematicians like Le Verrier counseled restraint, surmising that "after thirty or forty years of observing the new planet we will be able to use it in turn for the discovery of the one that follows it in order of distance from the Sun."

FALSE ALARMS

There was one other possibility open to those eager to snatch the planet-discovery laurels in a hurry—to look more closely at the orbit of Uranus, searching for perturbations that could not conveniently be explained by the now-known effects of Neptune and the other known planets. Uranus orbited faster, appeared brighter, and had been observed longer than Neptune, so that even though the effects of a far-out planet would be small, the perturbations might, with diligence, be teased out of the available data.

Between 1850 and the early 1900s, therefore, several astronomers gave this method a try. In 1877, young David Todd, who was to become a professor at Amherst College, analyzed the orbit of Uranus to calculate the position of the unknown planet at 52 astronomical units from the Sun, nearly twice the distance of Neptune. But though he searched for 30 nights using the 26-inch telescope at the U.S. Naval Observatory in Washington, DC, he found no trace of the object. In 1892, Isaac Roberts, a wealthy British amateur who made pioneering photos of the heavens, used his photographic skills to search for two planets that had been predicted a decade earlier by astronomer George Forbes at 100 and 300 astronomical units from the Sun. Again, nothing was found. In the early 1900s, Harvard astronomer William Henry Pickering made a virtual industry of predicting planets at the edges of the solar system. Between 1908 and 1932, he published papers identifying seven undiscovered planets, which he gave the provisional names "O," "P," "Q," "R," "S," "T," and "U," in sequence following the "N" of Neptune. His predictions proved no more fruitful than any of those in the past.

The available data, judging from the wide variation in predictions, seemed to give no conclusive answer to the whereabouts of a supposed planet. The only thing on which astronomers agreed was that telescopic searches for new planets in the outer solar system had turned up nothing.

A BRAHMIN TAKES UP THE QUEST

With such discouraging results, only a person with remarkable—perhaps unwarranted—confidence would take up the search for a new planet. Such a person was Percival Lowell. Born in 1855 to a wealthy and venerable Boston family that had already produced notable judges, industrialists, and literary figures, he grew up surrounded by overachievers. His father was a prosperous textile mill owner—Lowell, Massachusetts, bears the family name—and his siblings included a brother, A. Lawrence Lowell, who became president of Harvard University, and a sister, Amy, who was to become a freewheeling avant-garde poet in the early 1900s.

Percival, as the oldest boy in the family, was expected to do great things. His parents took the children for several years of travel in Europe, where young Lowell picked up a number of languages and received a liberal education in the arts and sciences. But he had a penchant for the exotic that appeared, from his youngest days, to direct him toward unconventional pursuits. Chief among them was astronomy. "Donati's comet of 1858," Lowell claimed, "being my earliest recollection . . . I can see yet a small boy half way up a turning staircase gazing with all his soul at the evening sky." Later, his brother recalled, Lowell eagerly read books on astronomy and often observed the heavens from the roof of the family house in Brookline, Massachusetts, using his own small telescope.

During four years at Harvard, Lowell continued his scientific pursuits, taking courses from Benjamin Peirce, a mathematician who had published work on the motions of Uranus and Neptune. Just as he had in his precollege days, Lowell achieved notable honors in all academic studies, and Peirce is said to have regarded him as one of the most brilliant intellects he had ever met. However, following graduation and the obligatory year as a footloose traveler in Europe, Lowell turned his efforts to more mundane affairs and for a while seemed to have settled into a typical Lowellian career in the family business. He entered the office of his grandfather, John A. Lowell, and worked as a mill manager, living the life of a wealthy young man-about-town, while his bank account swelled to healthy proportions.

After six years of this, freed by financial security to pursue his deep interest in new and exotic ideas, Lowell struck out for uncharted territory. In 1883, he moved to Tokyo where he quickly picked up the language and, over the next 10 years, became a noted correspondent and popular writer about Japan and Korea, riding a wave of enthusiasm for things Oriental that was sweeping across America at the time. His books, the most notable *The Soul of the Far East*, published in 1888, established his reputation as a gifted communicator in New England literary circles.

Still, Lowell retained his childhood fascination with the heavens. He was already familiar with the work of Giovanni Schiaparelli, an Italian astronomer who was known for his observations of Mars. Schiaparelli claimed that Mars was covered with a spiderweb of lines—he called them *canali*, or channels—which he supposed to be a network of interlocking natural watercourses, connecting the icy polar caps with dark regions that he identified as seas. To Schiaparelli, Mars was a sort of large-scale Venice, with the canali turning the surface of the planet into a geometric patchwork of islands.

Lowell was fascinated with this idea, speculating that Schiaparelli's canali might actually be canals, the handiwork of an intelligent race of Martians. But he knew that the idea was likely to stir controversy and that sustained observation and careful scientific record keeping was necessary to follow up Schiaparelli's seminal work. So when he heard, in 1893, that Schiaparelli's eyesight was failing and that he was essentially retiring from astronomy, Lowell decided to take up the study of Mars on his own.

Just has he had become an expert on the exotic Far East, Lowell set about systematically and energetically to establish himself as an expert observer of Mars. Though he already possessed some excellent telescopes, he knew that the weather and viewing conditions in the Boston area were not suited to a careful scrutiny of the planet. He chose a site near Flagstaff, at an altitude of 7,200 feet in the highlands of northern Arizona, as the ideal location for a planetary observatory. The clarity and darkness of the skies were unmatched, and the town was right on the main southern transcontinental railway line. Lowell could board a train in Back Bay and debark at his western observatory a few days later.

Over the next two decades, the Lowell Observatory became a well-known site for astronomical research in the United States, partly because of the excellence of its facilities and also because of the skill of the astronomers Lowell hired to work with him. Lowell and his staff conducted nightly observations of the planets, recording the changing appearance of Mars (and other planets as well) and publishing bulletins and scientific journal articles on their results.

Professional astronomers, however, regarded Lowell with a generous measure of skepticism. It was not simply that he was seen as a wealthy dilettante. Very few people were paid to do science up until the 20th century, and so a large fraction of the astronomical community was composed of amateurs, most of them wealthy like Lowell, who had state-of-the-art telescopes and who contributed to technical journals. Rather it was Lowell's writings for the general public, filled with extravagant claims, that rankled his more conservative scientific colleagues. With his reputation as a popular writer on the Far East already established, Lowell began to write popular pieces on astronomy for magazines and, in 1895, published a first popular book on Mars, complete with delicate color drawings of the planet. This was followed by a series of more provocative books, setting forth his ideas about what he regarded as the overwhelming evidence that the red planet was inhabited. In the first of these, *Mars and Its Canals*, published in 1906, Lowell boldly claimed that Mars possessed life—not just plants but also life of "a higher and more immediately appealing kind." "That Mars is inhabited," he proclaimed, "we may consider as certain."

The public ate this up, but astronomers grumbled that Lowell was presenting rank speculation as well-established fact. Many seasoned observers, like Edward Emerson Barnard of Yerkes Observatory, doubted the trustworthiness of his maps and thought that Lowell was claiming to see things through the telescope that were simply not there. Others questioned even the possibility of life on Mars, among them Alfred Russel Wallace, the noted naturalist who shared credit for the theory of natural selection with Charles Darwin. In 1907, Wallace devoted a small book, *Is Mars Habitable?*, to rebutting Lowell, arguing that Mars was too cold to have rivers, canals, or seas and that the features Lowell was mapping were natural landforms produced by normal geological processes. "I venture to think that his merit as one of our first astronomical observers will in no way be diminished by the rejection of his theory," Wallace concluded.

Many space missions to Mars in the past few decades have proven Wallace essentially correct. The red planet is cold and dry, and though there are many features on its surface that were shaped by running water some time in the distant past, the canals that Lowell mapped so meticulously do not exist. Lowell's canals, it seems in retrospect, were creations of his active mind, patterns drawn by connecting the dots on the planet that he could barely glimpse through the roiling of the Earth's atmosphere. Much of what Lowell wrote about Mars, both in his popular books and in the publications of his observatory, now seems as quaint and unsubstantiated as the alchemical fantasies of medieval scholars.

Yet Lowell's observatory, from its founding in 1894, established itself as an active and productive scientific institution, and it remains a premier astronomical research center to this day. If Lowell's obsession with life on Mars tarnished his professional reputation, he still set high standards for systematic record keeping and careful observation; he hired some of the best astronomers in the business to work with him. His own work included observations of other planets, and, as a trained mathematician, he devoted his studies not just to making sketches, but to highly technical explorations of the physics of the solar system as well. He may have been faulted for overzealousness, but he was not a pseudoscientist or a crackpot.

It was Lowell's mathematical training, along with his predilection for investigating "big questions" at the frontiers of science, that led him to take up the search for a planet beyond Neptune. More than half a century had passed since the discovery of that planet, and to Lowell, it seemed that the time was right for a thorough search for the supposed planet beyond, which he called "Planet X." Lowell had been aware of the problem since his student days with Benjamin Peirce at Harvard, and it seemed to him that, given the fact that Neptune had not yet completed a half turn around the Sun, the best hope for predicting the position of Planet X lay in looking for its effect on the orbit of Uranus, as others had done before him. Lowell believed he could succeed where others had failed because he had a longer timeline of observations of Uranus than did previous astronomers, the money to hire assistants to help with the laborious computations required to turn the observations into predicted positions of Planet X, and some of the best telescopes in the world with which to conduct a search.

And so, in 1905, Lowell took some time from his busy schedule to mount a quest for a trans-Neptunian planet. At the outset, he realized that it would be a difficult task, as he wrote in a later memoir on the search: "For so complicated is the problem that all elementary means of dealing with it lead only to error. The sole road to any hope of capture lies through the methodical approach of laborious analysis."

To begin his methodical quest for Planet X, Lowell contracted the services a of a mathematician, William Carrigan, who worked at the Naval Observatory in Washington, DC, followed a few years later by a young MIT graduate, Katherine Langdon Williams, who worked in Lowell's Boston office. For the next 10 years, these human "computers" provided Lowell with predicted positions of Planet X. They employed elaborate methods, similar to those that had been used to find Neptune, beginning with the assumption that the average distance of Planet X from the Sun would obey the Titius-Bode relation—in other words, it would be about 47.5 astronomical units from the Sun. Lowell guided the mathematical details of the work, with his personal involvement in the project waxing and waning as other commitments intervened.

Meanwhile, Lowell directed astronomers in Flagstaff to take photographs of regions of the sky where the predicted planet might be hiding. In the early years of the search, before the mathematical calculations had progressed very far, the Flagstaff astronomers simply pointed their telescopes along the plane of the solar system, looking for a faint dot of light that moved from night to night among the stationary pinpoints of the stars on the photographs. Lowell knew that the first random searches were unlikely to be successful, but he expected that as his predictions improved, he'd be able to zero in on his target. The early stages of the search were practice runs, important for developing effective strategies for flushing out the elusive quarry.

There were a lot of kinks to be worked out. The first photos, 1-hour exposures taken with Lowell's big 24-inch refractor, revealed nothing of interest and brought to light a serious problem: the field of view of the telescope was so narrow that it would take ages cover the area where a suspected planet might lie. Lowell obtained a smaller telescope with a wider coverage of the sky and continued the search, again with no success. It was clear that even better equipment was needed, along with better predictions to narrow the area to be searched.

So Lowell continued to make improvements. In 1911, he purchased from the German optics firm Zeiss a "blink comparator," a microscope that allowed astronomers to flip back and forth between two photographs while viewing them, making it easier to spot objects that appeared to move from one image to the next. He tweaked his computations; he borrowed a larger, wide-field telescope from Sproul Observatory at Swarthmore College and moved it to Flagstaff; he pestered his observers for more observations and deeper exposures. Perhaps Planet X was less reflective or smaller than expected and consequently fainter than his telescopes could detect. But by September 1915, when Lowell published a lengthy mathematical study summarizing his work on Planet X, *Memoir on a Trans-Neptunian Planet*, he had to concede that the quest was more difficult than he had anticipated: "Owing to the inexactitude of our data, then, we cannot regard our results with the complacency of completeness we would like." Modest words from a man used to forging ahead with patrician confidence. Lowell, according to the testimony of friends and colleagues, was frustrated and disappointed that Planet X remained undetected after so much work.

Overwork and stress took their toll. Percival Lowell's personal quest for Planet X was cut short suddenly on November 12, 1916, when he died in Flagstaff of a massive stroke. He was laid to rest under a dome of blue glass, set among the domes of working telescopes on Mars Hill, the site of his Flagstaff observatory. Though he never realized his dream of joining the ranks of the great planet finders Herschel, Adams, and Le Verrier, he left behind a first-class astronomical research facility, a staff of experienced astronomers, and a legacy—both financial and intellectual—that was to energize the next round of solar-system discovery.

HELP FROM A KANSAS FARM BOY

For a decade after Lowell's death, the search for Planet X floundered. Though Lowell had specifically provided more than a million dollars to continue research in Flagstaff, the terms of his will were immediately contested by his widow, Constance, and Lowell

Observatory operated in more or less financial limbo until the affair was settled.

In 1927, with the legacy decided at last, Roger Lowell Putnam, Lowell's nephew, took over as trustee of the observatory, clear in his mandate to carry on where Lowell had left off. After discussing matters with Lowell's successor as director of the observatory, Vesto Slipher, and other astronomers on the staff, Putnam decided that the search for Planet X would be continued, this time with a larger telescope designed specifically for planet hunting. He persuaded Percival's brother Lawrence to help out by supplementing the observatory's funds for the project and ordered a 13-inch lens with a wide field of view from the distinguished firm of Alvan Clark and Sons in Cambridgeport, just across the river from Boston. A telescope mounting for the lens and a dome for the telescope were constructed up the hill from the Lowell 24-inch dome in Flagstaff. The new telescope went into operation in February of 1929.

But who was to carry out the night-to-night routine of photographing and measuring? Vesto Slipher, as the man in charge of operations in Flagstaff, knew that the search for Planet X would be at best difficult and time-consuming, and at worst unproductive. He and his staff already had full schedules, and it would not be easy to hire a university-trained astronomer to take on such a job. It was not a task for an ambitious professional, eager to make a splash in academic circles, since there was a good possibility that the search would fail. The job called, rather, for someone with stamina and focus, able to derive pleasure from the simple act of doing a difficult job well.

In 1928, Slipher received a letter from an amateur astronomer from Kansas, a young man named Clyde Tombaugh, enclosing some drawings of Jupiter that he had made with a homemade telescope. The drawings "look fairly good for such a chap working all alone," Slipher confided in a colleague. "We are wondering if he might not be able to make exposures with our new 13-inch camera." Tombaugh, it turned out, was not only able, but willing, and in a few weeks, he and Slipher had settled on a deal: Tombaugh would come out on a trial basis for a few months. He'd see if he liked the work, and the observatory would see how good he was at doing what needed to be done.

At the time he set out for Arizona, Clyde Tombaugh was 22 years of age and an unlikely successor to Percival Lowell. Raised on a farm near Burdett, Kansas, a tiny community in the west central part of the state, he had graduated from high school in 1925 with the ambition to become a professor but stayed around to help out the family; times were hard, and there was no money to pay for a college education. Tombaugh shared an interest in astronomy with his Uncle Lee, who had bought him a telescope from the Sears catalog several years earlier, but farming left little time for casual stargazing. Still, though he spent his summers in the fields, he found time during the slack winter months to construct several telescopes, gradually improving his technique. In 1928, when a hailstorm destroyed the family crops just before harvest, young Tombaugh began thinking of alternatives to rural life. He entertained the notion of working on the railroad or perhaps going into business as a telescope maker. And just to satisfy himself that he really was doing decent work with his telescope, he sent some of his sketches to Slipher in Flagstaff, choosing Lowell Observatory because, as he later wrote, "it was the only planetary observatory I knew of." It was a choice that not only changed his life, but the course of planetary astronomy as well.

On January 15, 1929, Clyde Tombaugh stepped down from a westbound Santa Fe Railroad train at the Flagstaff depot where Vesto Slipher was waiting to drive him up to his quarters on Mars Hill. Though the new telescope had not yet arrived, Tombaugh immediately displayed an aptitude for the long hours and tedious work of a night observer. "He has a good attitude," a relieved Slipher reported to Roger Putnam, "careful with apparatus, willing to do anything to make himself useful, and is enthusiastic about learning and wants to do observing." Tombaugh, recalling those first weeks, commented with Midwestern reserve, "The work, most certainly was different from that on a Kansas farm."

And so the search for Planet X began again. Though Lowell's mathematical predictions had been an inspiration for the project, Tombaugh didn't trust them—Lowell had been wrong in the past, and he'd changed his predictions over the years as he'd tweaked his calculations. The young observer decided to troll for the planet rather than cast at an uncertain target, and so he began systematically surveying the region around the zodiac, the band across the sky through which all the known planets move as they orbit the Sun. By April 1929, with the new 13-inch telescope up and running smoothly, Tombaugh began what was to become a nightly routine. He would point the telescope at patches of sky 180 degrees from the Sun (the so-called "opposition point") and begin taking a series of hour-long exposures. Weather permitting (and Flagstaff skies are usually clear except in the months of July and August), he would take as many photos as he could between dusk and dawn. He would develop the photographs—which were 14 × 17-inch glass plates—and then set them aside to be scrutinized during the daytime.

The measuring process was as time-consuming as the observing. Tombaugh would slide a pair of photographs of the same area taken on successive nights into the twin plate holders of the Zeiss blink comparator. Looking through the eyepiece of the comparator, Tombaugh would adjust the positions of the plates so that the stars on the two photographs aligned with one another. Then, by flipping a lever, he could view one plate, then the other. Distant stars, which do not appear to move from night to night, would not change position as the lever was flipped; a body in orbit around the Sun, however, being much closer than the stars, would move slightly across the starry background, making itself obvious. Only a small portion of each plate could be seen through the comparator eyepiece, so Tombaugh had to systematically scan until the entire area of the plate had been examined. It was like plowing the furrows on a 40-acre field.

Each night's observations turned up plenty of objects that moved. Some of these were false alarms—scratches on the photographs could be easily mistaken for moving objects. The trick in weeding them out was to blink a third image of the same region. Only a real object would appear in all three photos. Even real objects, though, could fool you. There were many thousands of asteroids, most between the orbits of Mars and Jupiter, that showed up on the images. Tombaugh learned to recognize these by their relatively large motion—asteroids moved about 7 millimeters a day on the photographs he took. Planet X, being much farther away, was expected to move a millimeter or less.

Tombaugh was conscientious about his work, but it was a strain. There were so many stars on the plates that it took a long time, carefully flipping the blink lever back and forth, to completely inspect a pair of photographs. And it took a lot of patience. The task got especially tedious when the opposition point for a particular evening was located in the Milky Way. Some of the plates, Tombaugh estimated, contained images of more than 300,000 stars. It was impossible to keep with the work. Since he had to observe on clear nights, Tombaugh could only inspect the plates on days when cloudy weather intervened or at times of the month when the full moon made long exposure photography impossible. There were times, that first year, when Tombaugh was faced with a growing stack of unexamined plates.

Scarcely a year after his arrival in Flagstaff, the young observer from Kansas hit pay dirt. It was, he recalled, the afternoon of February 18, 1930, and Tombaugh was catching up on the measurement of three plates taken in the direction of the constellation Gemini on January 21, 23, and 29 of that year. Suddenly, as he moved the microscope around to inspect the plates—all the while blinking back and forth between the pair—he saw a faint dot jumping before his eyes. "The experience was an intense thrill," he recalled, "because the nature of the object was apparent at first sight. The shift in position between January 23 and 29 was about right for an object a billion miles beyond Neptune's orbit. . . . The images were quickly found on both plates of January 21." This was not an asteroid orbiting relatively nearby, and it was not a scratch or a mote of dust. It was, Tombaugh was certain, Planet X. He called Vesto Slipher and the other Lowell astronomers to check out his discovery. Lowell, interred a few hundred feet away, may have been smiling.

No one at Lowell Observatory, of course, was about to announce the discovery of a new planet without confirming the observations. To the frustration of the astronomers, Mars Hill was blanketed with clouds on the evening of February 18, but on the 19th, Tombaugh was able to photograph the Gemini region again with the 13-inch telescope. It had been three weeks since the discovery images had been taken, but the moving dot was visible, just where expected, on that night's plates. The next night, with the position confirmed, Lowell astronomers photographed it with their flagship telescope, the 24-inch refractor. Even at high magnification, the planet looked like a featureless dot, indicating that it was smaller than Uranus or Neptune. Lowell's last computations, published more than a decade earlier, called for an object noticeably larger and brighter, but the new planet was not far from his predicted position. And in any case, it was clearly way out there, orbiting beyond Neptune. Planet X, it seemed, had been found.

PLANET X BECOMES PLUTO

Although, as Vesto Slipher commented to Tombaugh at the time, "this could be very hot news," Lowell astronomers sat on their discovery for several weeks. "Don't tell anyone about the discovery," Slipher cautioned Tombaugh. "We need to keep it secret for a few weeks to study the object." They needed time to observe Planet X enough to make sure that it wasn't a comet, to determine its orbit as accurately as possible, and to firmly establish the observatory's priority in the matter. Once the word leaked out, Slipher knew, larger telescopes would take up the study in earnest, and he wanted to make sure that Lowell astronomers got a chance to find out all they could with the observatory's instruments before others got some of the glory. Tombaugh didn't even tell his parents; they found out that their son had found a new planet only when the editor of a local paper called them for comment.

News of the discovery of Planet X was released officially on March 13, 1930, the 75th anniversary of Percival Lowell's birth and, coincidentally, the 144th anniversary of William Herschel's discovery of Uranus. The next day, a bold headline in the *New York Times* announced "NINTH PLANET DISCOVERED ON EDGE OF SOLAR SYSTEM; FIRST FOUND IN 84 YEARS; LIES FAR BEYOND

NEPTUNE; Sighted Jan. 21 After 25 Years' Search Begun by Late Percival Lowell." Lowell was mentioned prominently in the article, and Tombaugh, who is remembered today as Pluto's discoverer, was given only passing mention as a "photographer at the observatory" who first noticed it. The new planet, it seemed, was getting a bit of spin from the Flagstaff astronomers, eager to give their founder credit for mathematically predicting its presence, just as Adams and Le Verrier had predicted Neptune. This, it turned out, was a bit premature, given that so little was known about it at the time. "Way out beyond Neptune," the *Times* article noted, "tagging bashfully behind its brothers, the new planet's exact whereabouts, size, and age are still unknown, and it has not even a name. . . . [U]ntil some one entitled to do so gives the sphere a name, it is to be known as "the trans-Neptunian planet.'"

Names were not long in coming. As the widow of the man who predicted the new planet, Constance Lowell weighed in with a number of recommendations. She first suggested Zeus, a logical choice, though one might argue that there already was a planet named after that god in its Roman form of Jupiter. Then, deciding that her late husband deserved a place on Mount Olympus, she put forward the name Percival. And if not that, then—her husband gracious and willing to defer to the fairer sex—perhaps Constance would be an even better choice.

Perhaps Mrs. Lowell was unaware of the fate of Galileo's Medicean Stars or Herschel's planet named George. Unsurprisingly, Mrs. Lowell's lobbying efforts were no more successful than those of her predecessors. The naming of planets, as we have seen in earlier chapters, was effectively decided in the court of common usage— textbook writers and professional astronomers simply began using names (invariably from mythology) that were less fraught with political baggage than the names of patrons or monarchs. Since the late 1800s, to be sure, discoverers of asteroids had been able to suggest names that were, as a rule, accepted by the astronomical community; there were just too many asteroids being discovered to worry about the proliferating list of names. Still there was no formal or informal procedure for full-fledged planets. Though a transnational organization of astronomers existed by this time—the International Astronomical Union had been founded in 1919—neither the IAU nor anyone else had any experience in the high-stakes game of planetary naming.

As head of Lowell Observatory, however, Vesto Slipher had some authority in the matter. During the spring of 1930, his desk was flooded with mailed-in suggestions for naming Planet X. Tombaugh got one letter from a young couple asking that the planet bear the name of their child. Mythological names, which had won by default in the past, were among the most popular: Odin, Persephone, Hercules, Hera, Pax, and Icarus. Slipher initially favored the name Minerva, goddess of wisdom, though it had already been assigned to an asteroid.

Cronus, the son of Uranus, at first seemed an even more appealing appellation, but that name had been submitted by an American astronomer named T. J. J. See who was persona non grata among his colleagues. See, a Missourian who held a degree from the University of Berlin, had begun his career as a promising observer of double stars. But he was insufferable, pompous, and sublimely assured of his intellectual superiority. He had worked at Lowell Observatory for two years, even serving as temporary director for a while, but he so infuriated the staff that he was summarily fired in July 1898 and eventually ended up working at various technical jobs for the Navy until his retirement in 1930. See was widely regarded by everyone except himself as a crackpot of the highest order. He'd published a bizarre book on the evolution of the cosmos, and a hagiography by a Mr. W. L. Webb, *Brief Biography and Popular Account of the Unparalleled Discoveries of T. J. J. See*, was widely believed to actually have been written by the astronomer himself. Cronos might be an apt name, Slipher acknowledged, but there was no way he would willingly encourage the mad astronomer by honoring his request.

In the end, Slipher was won over by a suggestion cabled by British astronomer Herbert Turner, who related a suggestion he had heard from a friend in Oxford, England. The gentleman's 11-year-old granddaughter, Venetia Burney, who had been studying mythology in school, thought that Pluto was an ideal name for a planet so far from the Sun. In Miss Burney's opinion, Pluto's realm was appropriately "murky and mysterious," just as Planet X must be. Slipher liked that, and the fact that the first two letters of Pluto were the initials of Percival Lowell didn't hurt. On May 1, 1930, in letter on Lowell Observatory letterhead addressed to the American Astronomical Society and the Royal Astronomical Society, he suggested that Pluto be the official name and that ♇, formed from Lowell's initials, be used as a symbol for the planet.

BUT IS PLUTO PLANET X?

Thus it was that the trans-Neptunian object discovered by Clyde Tombaugh became known as Pluto, the ninth planet of the solar system. Aside from the Lowell letter, to be sure, there was nothing legally binding anyone to use the "official" name, simply because there was no official body with the undisputed authority to name it. Yet people liked calling it Pluto, and though its discoverer, Clyde Tombaugh, never gave it his full blessing, it quickly passed into popular usage.

Yet while the name Pluto was widely accepted, the status of Pluto as the ninth planet of the solar system immediately came into question. Was it, in fact, the unseen body whose presence had been predicted by Percival Lowell's analysis of perturbations affecting the orbit of Uranus? Tombaugh, lacking confidence in the certainty of Lowell's mathematical approach, had opted for a systematic survey of the sky rather than a pinpoint search only near Lowell's target areas, and in later writings, he asserted that Lowell's calculation played no role in his discovery.

Tombaugh had to admit that Lowell's predicted position was only about 6 degrees from where Pluto was eventually found, but its average distance from the Sun was 40 astronomical units, about 5 astronomical units closer than Lowell had predicted. Lowell had gotten the direction right, in other words, but not the distance, so perhaps it was just a happy accident that the planet happened to be in the right place at the right time. In the months that followed the announcement of the discovery, astronomers worked to refine the orbit of Pluto through further observations of its position. At first, Lowell astronomers held their observations close to their chest, releasing information on the new planet's positions reluctantly, so that they would continue to have an advantage in characterizing its behavior. But by summer of 1930, there was a growing community of astronomers around the world who were collecting and analyzing data on the new planet.

Astronomers not only photographed the planet whenever possible, they also searched through archives of plates to see if it had been photographed, but not recognized, in the years before its discovery. Observers found Pluto on plates taken at Mount Wilson Observatory in 1919 and at Yerkes Observatory in 1921 and even on some plates taken during the earlier Lowell Observatory search in 1915, shortly before Lowell's death. With a longer timeline of observations at hand, it became clear that the predicted characteristics of Planet X bore only passing similarity to the real Pluto.

Lowell's last calculations had led him to expect a planet about 6.6 times the mass of the Earth orbiting about 43 astronomical units from the Sun. He expected the planet to be large enough to be seen as a distinct globe through a large telescope. In actuality, Pluto seemed to be considerably less massive than the Earth and so small that it appeared as an unresolved point of light. It orbited at about 40 astronomical units from the Sun, not 43; its orbit was more elliptical than expected; and the plane of its orbit was much more highly inclined to the plane of the solar system. William Pickering, who had predicted a bunch of planets in the past, was eager to claim that Tombaugh had discovered his Planet O, not Lowell's Planet X. But most astronomers, save a few strong partisans in Flagstaff perhaps, were skeptical of all such claims, preferring to regard Pluto as an odd duck, whose true nature remained to be seen.

Tombaugh himself was willing to grant that more work needed to be done. If Pluto was as low in mass as it seemed to be, then perhaps the irregularities that Lowell and other astronomers had detected in the orbits of the planets were the result of the influence of an as yet undiscovered planet—perhaps the "real" Planet X had yet to step forward. Accordingly, Tombaugh continued to work with the 13-inch telescope for over a decade, surveying, by his estimation, three-quarters of the entire sky visible from Flagstaff and examining over 90 million individual star images through the eyepiece of the blink comparator. Although he discovered a new comet, hundreds of new asteroids, and more than a thousand variable stars before he finally abandoned the search, he never found anything to rival Pluto.

Over the next three-quarters of a century, while schoolchildren around the world were learning that Pluto was outermost of the planets, astronomers were gathering more and more evidence on how different it was from the other eight. The story of how the visionary Lowell and the industrious Tombaugh found a new planet began to take on mythic proportions in the popular mind. But, like most heroic myths, it was just too good to be true.

CHAPTER 8

The Exploration
of Pluto

For everyone, as I think, must see that astronomy compels the
soul to look upwards and leads us from this world to another.

—Plato, *The Republic*, Book VII, 1.529, 380 B.C.

Astronomers have struggled to understand the nature of Pluto over
the eight decades since Clyde Tombaugh discovered it in 1930. They
had to await the development of powerful new telescopes on Earth,
aboard an aircraft, and in space, which enabled the investigation of
the dim, distant planet. More discoveries about Pluto flowed from
the application of new technologies to study it as telescopes were
equipped with more sensitive instruments.

The element of luck entered the campaign to understand Pluto
when an alert astronomer spotted something suspicious that other
scientists had ignored. And rare celestial alignments occurred,

enabling researchers to glean unprecedented amounts of crucial information.

As time went on after the discovery of Pluto, the planet advanced closer and closer to the Earth and Sun, following its 248-year orbit. In February 1979, Pluto swung within the orbit of Neptune, making it the eighth planet from the Sun for 20 years (if you count Pluto as a planet). In 1989, Pluto was at perihelion, its closest approach to the Sun, and it headed back out toward the distant reaches of the solar system, passing outside Neptune's orbit in 1999. In the last decade of the 20th century, astronomers began to discover other bodies beyond Neptune and sometimes beyond Pluto as well that revealed the existence of a predicted region of small icy bodies, the Kuiper Belt.

SIZING UP PLUTO

When William Herschel first sighted Uranus in 1781, he could immediately distinguish it from stars that he saw through the same telescope (see Chapter 3). The star images were pinpoints of light, but the telescopic image of Uranus had a perceptible shape and size. The Uranus image was round; astronomers refer to an image like that as a disk. Twentieth-century astronomers began trying to see the disk of Pluto so they could measure it. Some thought that Pluto might be as large as Earth or even larger. But when the attempts to measure Pluto began, they were foiled, because it is smaller than they imagined and their telescopes were inadequate to show the disk.

The largest and most powerful telescope that the world had ever known was completed in 1949 on Palomar Mountain near San Diego, California. If any telescope was going to reveal the disk of Pluto, surely the 200-inch Hale reflector was it. In March 1950, the necessary observations were made with it. The observer was Gerard P. Kuiper of Yerkes Observatory and McDonald Observatory, the most important planetary astronomer (not that there were many such experts at the time) in the United States. Later Kuiper would go on to found and direct the Lunar and Planetary Laboratory at the University of Arizona in Tucson. After his death, NASA named its largest airborne telescope facility the Kuiper Airborne Observatory in his honor. The KAO, as it was known, would be the very aircraft that flew across the sea to observe an occultation of a dim star by Pluto in 1988, as described later in this chapter.

On March 21, 1950, Gerard Kuiper observed Pluto at the prime focus of the 200-inch telescope, where he had mounted a disk meter, a device that produced artificial, illuminated disk images of known sizes. Riding inside the big telescope in an observer cage, Kuiper carefully compared the image of Pluto formed by the telescope at a magnifying power of 1,140 with various-sized disks. He found that Pluto's disk was 0.23 arc seconds in diameter. (For comparison, there are 3,600 arc seconds in one degree on the sky, and the Moon is about a half-degree in diameter.) Kuiper wrote that this measurement meant that Pluto's "diameter is 0.46 times the earth, midway between Mars's and Mercury's" diameters. A leading astronomer at Palomar Observatory, Milton Humason, tried his hand at the telescope with Kuiper's disk meter and got the same result. So Pluto was apparently the second-smallest planet, larger only than Mercury. We now know that the measurements were wrong and that Pluto is much smaller than Mercury. An accurate diameter measurement was not to come for decades after Kuiper's work, despite much trying. In the meantime, estimates of the diameter of Pluto by astronomers trying new methods kept getting smaller and smaller.

As years went by after Kuiper's attempt to gauge the disk of Pluto, other astronomers made new observations, which showed that Kuiper was wrong and that Pluto must be much smaller than he estimated. Some experts refined the knowledge of Neptune's orbit, finding that the lack of perceptible effect of Pluto's gravity on Neptune meant that Pluto's mass must be smaller than was previously believed. Over the years, estimates of the mass kept dropping. In 1980, two physicists prepared a graph of the estimated masses of Pluto versus the years in which the estimates had been made. Fitting a mathematical curve through the data points, they jokingly concluded that Pluto would soon reach zero mass. But the little planet did not vanish.

The first real advance in knowledge of Pluto since its discovery came in 1959, when Robert Hardie and Merle Walker used a 42-inch telescope at Lowell Observatory and a photomultiplier tube to repeatedly measure the brightness of Pluto. They found that the brightness varied periodically, as though the planet rotated once every 6.39 Earth days, much slower than any of the planets from Earth through Neptune but not as slow as Mercury or Venus. Apparently there were large light and dark areas on Pluto, which caused it to reflect alternately a little more or less of the Sun's light toward the observers on Earth as Pluto turned.

The decade of the 1970s opened with the measurement of Pluto's color. It has a reddish cast. Within that decade, astronomers using infrared sensors discovered that Pluto is coated with frozen methane (methane ice). Methane (chemical formula CH_4) occurs naturally on Earth as a gas in the atmosphere. It's most commonly encountered as natural gas, traveling through a pipeline to heat your home or liquified by compression to travel in bulk on huge tankers at sea. Fresh methane is produced by natural processes on Earth, including most famously by flatulence in cows. But you won't see frozen methane on Earth, except in a physics or chemistry lab. In contrast, there is much methane gas in the atmospheres of Jupiter and Saturn. Farther from the Sun, in the frigid atmospheres of Uranus and Neptune, tiny crystals of methane ice float in high-altitude clouds. On Pluto, the first methane found was ice on the ground.

PLUTO'S PARTNER

Until June 22, 1978, James W. Christy, a man without an advanced degree (most research astronomers have PhDs), was a little-known astronomer at the U.S. Naval Observatory (USNO) in Washington, DC, but by the end of the day, he was on his way to becoming world famous. It began as a routine day when Christy sat down to measure the position of Pluto among the background stars on a collection of photographic plates. The photographs were made a few months earlier with a 60-inch telescope at the USNO's Flagstaff Station near Lowell Observatory. The purpose of the position measurements was to provide new data on Pluto's motion through space so that its orbit could be calculated more precisely than before, and to get the best results, Christy had to look at the images of Pluto through a microscope.

As later reported by *Comet News Service*, "To his [Christy's] surprise, he immediately noted that the images of Pluto were elongated—all the exposures on 13 and 20 April showed a faint southerly extension, and those on 2 May showed a faint northerly extension." The "extensions" were in fact images of Pluto's previously unknown moon, now called Charon, at opposite ends of the moon's orbit. Charon is so close to Pluto (less than 11,000 miles) that at their great distance from the Earth, they could not be separately distinguished and were blurred together.

The discovery was soon verified on older plates at the USNO, where the extensions were present but had not been noticed before or had been noticed but dismissed as imperfections in the plates. Christy found that stars on the plates were perfectly round images, as Pluto's image should be, unless it had a moon that became briefly and barely visible at opposite ends of its orbit, where it was farthest from Pluto as seen from Earth. The "extensions" were not the images of one or more background stars, because other photos taken when Pluto was not at that position showed that there was no star in the right place. And the orbital period of the moon was the same as Pluto's rotation period, 6.39 days.

The USNO astronomers who had overlooked evidence for Charon before James Christy sat down to work were not the only ones. A photographic plate of Pluto, taken a year or two earlier than Christy's plates by an astronomer of NASA's Outer Planet Satellite project, showed the same telltale extension of Pluto's image that Christy later found, but it was not examined with sufficient care (if at all) at the time. As project director, J. Derral Mulholland of the University of Texas, Austin, later explained in *Science*, "Frankly we were not interested in Pluto, and the observations were taken primarily because they represented very little additional effort, and could eventually be useful to someone, somewhere, sometime."

Just as Clyde Tombaugh's discovery of Pluto is considered the greatest scientific feat ever accomplished at Lowell Observatory,

James Christy's finding of Pluto's first known moon was deemed a triumph of the USNO. To cite a discovery of comparable importance, USNO astronomers pointed back to 1877, when Phobos and Deimos, the moons of Mars, were found at their observatory. When arrangements for a press conference on the new Pluto moon became confused, with staff members directing journalists from one room to another in search of the briefing, one Naval Observatory astronomer apologized to a reporter for *Science News*, joking, "The last time we had a press conference here was 101 years ago."

Like Clyde Tombaugh, the Kansas farm boy who took his job at Lowell Observatory because, as he later told John Noble Wilford of the *New York Times*, "It beat pitching hay," Christy was an observatory assistant without strong academic credentials. And, like Tombaugh, who succeeded at a planet search when others expected that he would fail, Christy hit pay dirt. Many astronomers of the 1930s are now long forgotten, and in due course, many astronomers of the 1970s will be forgotten as well. But Christy's and Tombaugh's names will remain in the history books for centuries to come.

The discovery of Charon had a big scientific impact. It enabled Christy's supervisor, Robert S. Harrington, to calculate the orbit of Charon and to derive the first reasonably accurate value of the mass of Pluto. All earlier estimates were grossly in error—far too large. Pluto has only 1/418 the mass of Earth, far smaller than the mass of any of the eight previously known planets. But Pluto was still deemed a planet in 1978, and Harrington's calculations showed that Pluto and Charon could be considered the first double planet. This is because, as described in Chapter 2, in the cases of other planets and their moons, the center of mass of the planet-moon system is always inside the planet. But the center of mass of the Pluto-Charon system is located on the line between the two bodies in empty space. Pluto and Charon each orbit that point as it, in turn, orbits the Sun.

When we spoke with Robert Harrington (who has since died) a few months after he calculated the orbit of Charon for the first time, he mentioned that, as seen from Pluto at the time of our conversation, Charon would be north of the Sun in the sky. On Earth we see the Sun and Moon to be of equal size because, although the Sun in reality is about 400 times larger than the Moon, the Sun is also about 400 times farther away. However, Charon, although smaller than the Moon, is much closer to Pluto than the Moon is to the Earth. As a result, if you were to stand on the side of Pluto that faces Charon, Charon would look eight times bigger than the Moon does when seen from Earth. Harrington also said that starting in about 1983, Pluto and Charon would be oriented so that each of them would alternately pass between the other body and the Sun. With Charon looming as large as eight Moons in Pluto's sky, it would then eclipse not only the Sun, but also the three innermost planets, Mercury, Venus, and Earth, at the same time.

Scientists were much more interested in observing Pluto and Charon pass in front of each other as seen from Earth than in the vicarious thrill of imagining what an eclipse of the Sun would look like on Pluto. When two bodies alternately pass in front of and across each other, the events are called mutual occultations. By observing the light curve of an occultation—as the total light received from Pluto and Charon changed as one body was gradually blocked from view by the other and was then revealed again in full at the end of the event—astronomers could hope to make a crude map of the light and dark areas on Pluto, the existence of which Walker and Hardie had inferred in 1959, and perhaps also detect surface markings on Charon.

According to *Pluto and Charon: Ice Worlds on the Ragged Edge of the Solar System* by Alan Stern and Jacqueline Mitton, it was Leif Andersson, a young Swedish astronomer in Arizona, who first realized shortly after Charon was discovered that the mutual occultations would occur. Most importantly, Andersson calculated that mutual occultations of Charon and Pluto, which can be seen from Earth only for a period of a few years once every 124 years, were due to occur in the early or mid-1980s, so they were coming up soon and not decades in the future. Stern writes that "when planetary astronomers first learned of Andersson's prediction" of the upcoming occultations, "the salivating was reminiscent of Pavlov's dogs." This would be their big chance to learn more about Pluto.

DANCING IN THE DARK

The first reliable observation of one of the mutual occultations came on February 17, 1985, when Richard P. Binzel, a University of Texas graduate student, measured the change in brightness over two hours as Charon passed in front of Pluto. Like Walker and Hardie, Binzel (now a professor at MIT who served in 2006 on the IAU's Planet Definition Committee) used a photomultiplier tube, which was attached to a 36-inch telescope at McDonald Observatory near Fort Davis, Texas. A month previously, on January 16, astronomers at Palomar Observatory, where Kuiper had attempted to measure Pluto's diameter, obtained somewhat weaker evidence of a similar occultation of Pluto by Charon. Then on February 20, David J. Tholen of the University of Hawaii recorded an occultation of little Charon by Pluto with the 88-inch telescope at Mauna Kea Observatory on the Big Island of Hawaii. In observations of subsequent mutual occultations, much new information was gleaned, including a 1988 report by Binzel that "the data may be attributed to a direct detection of polar caps on Pluto."

At the same time that astronomers began planning to observe the mutual occultations, they also were hoping to observe occasional stellar occultations when Pluto would pass in front of a star. A stellar occultation resembles a total eclipse of the Sun in that the shadow that Pluto casts when it passes in front of a star is only visible along a thin track across the Earth. But while predictions of the path of totality of a solar eclipse can be made with high accuracy years in advance, forecasts of stellar occultations are much less precise. They can be thrown off if you don't know the exact diameter of the eclipsing object or don't know its orbit with sufficient precision. We certainly did not know Pluto's exact diameter very well in the 1980s; nor did we know its orbit with great precision (recall that this was the reason that Christy sat down to measure Pluto's position on his set of photographic plates in 1978).

Pluto had been predicted to occult a faint star on April 6, 1976. Remarkably, an occultation of the star was successfully observed as predicted, but it was not Pluto that eclipsed the star! As Pluto slowly approached the star (which was about five times brighter than the planet) it became likely from new measurements of the planet's track that Pluto would just miss passing in front of it. Yet Alistair R. Walker, an astronomer at the South African Astronomical Observatory, observed an occultation of the star that lasted 50 seconds. Charon had occulted the star. Walker's measurement of the event duration with a 40-inch telescope at Sutherland showed that the diameter of Charon was about 750 miles or possibly a bit more.

The data from the occultation of a star by Charon was helpful, but still astronomers wanted to measure a stellar occultation by Pluto. That would give them definitive evidence of whether Pluto had an atmosphere (as some suspected) or not. When there's no atmosphere, the star will drop out of sight in an instant when an occultation begins. But when there is an atmosphere, it bends the starlight in such a manner that the drop in brightness is prolonged, not instantaneous. June 9, 1988, was the date when just such an occultation was finally observed, after some missed opportunities because of bad weather or other problems. Now astronomers fanned out across the landmasses in the South Pacific and took to the air in NASA's Kuiper Airborne Observatory.

The observations were a great success: not only did the 12th-magnitude star not wink out as soon as Pluto passed in front of it, the star disappeared slowly, allowing accurate measurements of the effect by many telescopes. The shadow of Pluto swept across the Earth at nearly 40,000 miles per hour, according to an MIT publication, but the KAO intercepted it correctly, while flying at 41,000 feet over the Pacific, 500 miles south of Pago Pago. The results indicated that Pluto unquestionably has an atmosphere, extending hundreds of miles and perhaps much farther into space. *New Scientist* commented that "[m]any astronomers doubted whether Pluto was massive enough to hold an atmosphere," but now there was no doubt. Alan Stern, who was among those predicting an atmosphere of methane and perhaps other substances, wrote to us at the time that the new occultation data provided "incontrovertible evidence for gaseous methane" on Pluto and also confirmation of predictions by a few other experts and himself "of a hugely extended atmosphere—this is caused by Pluto's very low gravity."

The mutual occultations of Pluto and Charon were due to end in 1990, and a new series would not begin until 2109. But by 1989, astronomers were awaiting another source of unparalleled information on Pluto: the Hubble Space Telescope (HST). Launched in April 1990, the HST featured what had been billed as the most perfect telescope mirror ever made. Pluto had now rounded perihelion and was heading back out toward the far reaches of its orbit, but the HST was expected to image its surface features in finer detail than occultations by Charon ever could provide, and to gather new information on the composition of Pluto's icy surface. Four separate projects to study Pluto with the HST were approved before the launch, including one headed by Alan Stern.

The combined data from the mutual occultations of Pluto and Charon and from the occultations of stars by each of the two bodies gave improved estimates of their diameters. The total mass of the Pluto-Charon system was known from its orbital properties. Planetary scientists used this information to gauge the nature of Pluto *beneath* its surface. Some earlier estimates were clearly wrong. For example, after the discovery of Charon, there were attempts to determine the density of Pluto. Investigators wanted to know if Pluto is hard as a rock, like Mars, or lighter than water, like Saturn. But inadequate information and perhaps questionable assumptions led them astray. Some had thought Pluto rocky, but others now estimated that it had a low density. Edwin C. Krupp of Griffith Observatory in Los Angeles, California, captured the implications of that conclusion in 1978 when he joked that "Pluto is no longer the small, rocky planet that didn't fit the pattern of the solar system, it is now the small, fluffy planet that doesn't fit the pattern of the solar system." This was aptly put, but, as we later learned, it was wrong.

Observations of the mutual occultations also showed that Pluto has a higher albedo than Charon, meaning that it is more reflective of visible light. While Pluto is coated with methane ice, Charon's surface, most likely, is frozen water. Richard Binzel wrote in 1990, "To human eyes, Charon would appear a bland, neutral gray, and Pluto would be reddish." Crude maps of the surfaces of the two bodies showed not only that Pluto is brighter at the poles than at the equator, but that its south polar cap reflects about 98 percent of incoming sunlight, like a bright Arctic vista on Earth, while the darkest spots on Charon are black as coal. It seemed at the time that Charon had no detectable atmosphere, while Pluto had an extensive atmosphere of methane. There might be other gases on Pluto, experts believed, but they had not been found. As Pluto's highly elliptical orbit takes it from a distance of 29.5 astronomical units from the Sun at perihelion to 49.4 AU at aphelion, gaseous methane probably freezes and drops to Pluto's surface like snow, a phenomenon first suggested by Alan Stern and his collaborators. (One AU is about 93 million miles.) A fresh coating of frozen methane, deposited once every 248-year Pluto orbit, may explain why Pluto's surface is more reflective than Charon's and in fact is seven times more reflective than our Moon's.

By 1989, with much of the occultation data in hand, planetary scientists were in a good position to investigate the Pluto interior. They calculated various possible physical models for Pluto to see which ones best fit the data, with fewest assumptions. Two of the experts, Steve Mueller of Southern Methodist University in Dallas and William B. McKinnon of Washington University in St. Louis, calculated that Pluto is "probably about 75 percent rock and 25 percent ice by mass," as they explained in the magazine *The World & I.* The rock, they asserted, makes up a large core of the planet, surrounded by 100 to 200 miles of frozen water that is topped with a few miles of methane ice and perhaps other frozen substances.

Shortly after the Hubble Space Telescope was launched in April 1990, NASA discovered that the much-vaunted main mirror had been incorrectly shaped. The error was very small, but it meant that the telescope was not in perfect focus. It would not be repaired until late 1993, and meanwhile Pluto experts had to rely on other telescopes and instruments for new advances. One such advance came in May 1992 at the 150-inch United Kingdom Infrared Telescope (UKIRT) on Mauna Kea, which was equipped with a powerful new spectrometer. The UKIRT team, led by Tobias C. Owen of the University of Hawaii, discovered the characteristic spectral signatures of frozen nitrogen and frozen carbon monoxide (CO). Calculations then indicated that nitrogen ice is actually the most common substance on the surface of Pluto, with CO ice a distant second (among detected materials) and methane ice even less prevalent. (The frozen methane had been found first because it is spectroscopically easiest to detect.)

When Pluto experts met in Flagstaff in July 1993, they disagreed about the size of Pluto. The very atmosphere whose discovery in 1988 had been hailed as a major advance in understanding was cast as the villain in the Pluto size debate. The atmosphere contains a haze layer, some said, while others pointed to possible temperature effects in the atmosphere. The atmosphere might cool off with altitude and then switch to warming with altitude above a certain height. The purported haze layer or the temperature changes or both could have complicated the interpretation of the stellar occultation measurements made in 1988 by the Kuiper Airborne Observatory and by telescopes on the ground. Pluto researchers needed a repaired Hubble Space Telescope, which would be capable of taking direct photographs that clearly separated the images of Pluto and Charon, enabling them to be accurately measured. In photos from ground-based telescopes, the two objects could be at best barely distinguished from one another. With a properly operating Hubble, the size measurements would be easy. To correctly understand the nature of Pluto's atmosphere and icy surface and their interactions with one another, astronomers needed to know the temperature of Pluto, but different telescopes gave answers that disagreed by about 20 degrees Fahrenheit and possibly more. One temperature estimate didn't fit Pluto's atmospheric properties, but was that measurement wrong or were the atmospheric data misinterpreted? It's no wonder that *Nature* headlined a report on the Charon-Pluto problems "An enigma orbiting a puzzle." The report's author, William B. McKinnon, stressed that exact information on Pluto's diameter was needed to help solve the problems, and he asserted that Pluto is "a critically important object, a large rock-ice world, and a planet by convention, born at the boundary between the realm of gas giants [the planets Jupiter through Neptune] and the Kuiper belt of comets beyond." Other astronomers might disagree with McKinnon's characterization of Pluto as "large" or as a planet, but none would doubt Pluto's continuing scientific importance.

HUBBLE TAKES A LOOK

Hubble alone would not be able to solve all the mysteries of Pluto. By 1993, Alan Stern was well into a campaign to convince NASA to launch a fast space probe to Pluto. He believed that seasonal changes occurring on Pluto as it receded from the Sun, a journey underway since 1989, would make it vital to reach and explore Pluto prior to 2020. The next perihelion will not occur until 2237, so the space mission was urgent. Stern also concluded from radio telescope measurements in 1991 and 1993 that "Pluto's atmosphere is dominated by nitrogen or carbon monoxide rather than methane." So, just like in the surface ices of Pluto, methane was present but not the prime component.

In 1990, even in its partly out-of-focus condition, Hubble obtained snapshots that clearly separated Pluto and Charon but left much to be desired. In December 1993, NASA astronauts aboard the Space Shuttle Endeavor repaired the telescope. This included installing corrective optical devices to fix the focus in some existing Hubble instruments, including the Faint Object Camera, and replacing the other on-board camera with a new one called Wide Field and Planetary Camera 2. They also installed new spacecraft systems to reduce jitter in the solar panels that was shaking the telescope and to help it point at and track targets. For some astronomers, Pluto was a prime target, and observations soon began. In May 1994, NASA released an image described as "the clearest view of the distant planet Pluto, and its moon Charon, as revealed by the Hubble Space Telescope." The picture had been obtained when Pluto was 2.6 billion miles from Earth and Pluto and Charon were "shown as clearly separate and sharp disks."

The Hubble-Pluto saga continued. By 1996, NASA was reporting that the orbiting telescope had pictured "nearly the entire surface of Pluto." NASA revealed that "[f]or the first time since Pluto's discovery 66 years ago, astronomers have at last directly seen details on the surface of the solar system's farthest known planet." The bright and dark areas mapped by the Hubble might indicate topographic features of the planet, or they might represent variations in the distribution of frozen gases on Pluto's surface, or both. The images were obtained by a science team composed of Alan Stern, Marc Buie, and Lawrence Trafton. Buie, a veteran Pluto researcher at the Southwest Research Institute in Boulder, Colorado, said that Pluto was now "a world which we can begin to map and watch for surface changes." Stern called the lighter areas "as bright as fresh Colorado snow." The darker areas are perhaps discolored by chemical effects of ultraviolet light from the Sun and cosmic ray particles from interstellar space on the surface ices.

By 2005, astronauts had installed the powerful Advanced Camera for Surveys (ACS) on the Hubble, and on May 15, 2005, the ACS snapped pictures showing that Pluto and Charon had company. Twenty-seven years after James Christy discovered Charon, a team led by Stern and Harold Weaver, a planetary specialist at the Applied Physics Laboratory of Johns Hopkins University, found what appeared to be two much smaller moons, a few times farther from Pluto than Charon. In February 2006, the two little moons were confirmed with additional ACS observations. By then NASA was quoting Pluto's diameter as about 1,416 miles. The moons were named Nix and Hydra. They are the same color as Charon and are about 5,000 times fainter than Pluto.

PLUTO'S LONG SHADOW

There has been dramatic progress in studying Pluto and Charon in recent years thanks to a major increase in observed stellar occultations. Our knowledge of Pluto's orbit has improved so much that advance predictions of when Pluto will occult a star and where on Earth that event can been seen are now very reliable. Astronomers have been flocking to these events, sometimes shipping portable telescopes to locations on the ground track along which the occultation can be observed, or bringing specialized occultation instruments to mount on telescopes at existing observatories. When Marc

Buie told us about these expeditions in April 2009, he was about to fly to Africa to watch an occultation from Namibia and had shipped a 14-inch telescope ahead. He said that Plutophiles had been averaging about one occultation a year recently. The series of successful occultation observations began in 2002, after a hiatus of 14 years. The increased rate at which astronomers are observing occultations is because of the fact that, as seen from Earth, the position of Pluto against the starry background is moving closer to the plane of the Milky Way, where there are many more stars in a given area of the sky.

By 2002, when two stellar occultations by Pluto were observed, NASA had shut down the Kuiper Airborne Observatory after 21 years of service. Astronomers went to locations across South America for an event on July 20, 2002. They recorded the next occultation on August 21, 2002, with five telescopes at large observatories in Hawaii and California.

Pluto had been approaching perihelion on June 9, 1988, when a team led by James L. Elliott of the Massachusetts Institute of Technology had watched Pluto occult a star from the KAO, confirming that it has an atmosphere. In 2002, Elliott and colleagues wondered if Pluto's atmosphere might have frozen and collapsed as Pluto headed out away from the Sun after perihelion in 1989. Their observations on the two dates in 2002 revealed that Pluto's atmosphere indeed had changed since 1988. But, as Elliott and others wrote in *Nature*, the atmosphere had "expanded rather than collapsed." The August 21, 2002, observations indicated to the Elliott group and another observing team led by Bruno Sicardy of the Paris Observatory that the barometric pressure in the lower atmosphere of Pluto had doubled since 1988. The Sicardy group, also writing in *Nature*, called the change "a probable seasonal effect" and suggested that it may have occurred as the south polar cap of Pluto went into sunlight, causing nitrogen ice near the pole to sublime (change from solid to gas) and enter the atmosphere. A more speculative explanation was offered by Elliott, who wondered if there might be eruptions of some sort on Pluto that could replenish the atmosphere like vapors from a volcano.

The pressure increase on Pluto was also inferred from observations of the August 21, 2002, occultation by a team under Jay M. Pasachoff of Williams College in Williamstown, Massachusetts. They used telescopes on both the Mauna Kea and Haleakala volcanoes in Hawaii. The Pasachoff team detected sudden, brief increases in the apparent brightness of the occulted star as Pluto began to pass in front of it and again as the star emerged on the opposite side of Pluto. They attributed these brightness "spikes" to refraction effects (akin to the twinkling of stars as seen by a visual observer on Earth), perhaps because of an atmospheric disturbance on Pluto.

Twenty-five years after Alistair Walker first observed a stellar occultation by Charon, another event of that kind was observed by two large international teams, each using multiple telescopes in South America. One group of 13 astronomers brought special instruments called POETS (Portable Occultation, Eclipse, and Transit Systems) to mount on most of the telescopes that they used in Brazil (where clouds prevented successful observations) and Chile (where clear skies prevailed). POETS incorporated a high-speed digital camera, a computer, and a GPS receiver. According to a report by Amanda Gulbis (then with MIT) and the others, the occultation data ruled out the presence of any "significant atmosphere" on Charon. The second team, composed of over 40 scientists led by Sicardy, recorded the occultation with three telescopes, including the 323-inch Yepun reflector of the Very Large Telescope on Cerro Paranal in Chile. Both teams obtained similar values for the diameter (about 752 miles) and density (about 1.71 grams per cubic centimeter) of Charon. This confirms that Charon contains more rock than ice, but Pluto has a slightly larger rock content. The Gulbis team commented that the findings "seem to be consistent with collisional formation for the Pluto-Charon system," in accord with theoretical calculations made by William McKinnon at Washington University in 1989. Some theorists, like the late Ernst J. Öpik, an Estonian astronomer at Armagh Observatory in Northern Ireland, had argued that Charon formed by fission, splitting off from Pluto as it rotated. Öpik had also suggested in 1978 that "Charon may be a piece of bare rock with a density similar to that of the terrestrial planets or the Moon." The occultation teams confirmed, however, that it has a substantial ice content.

There was another occultation on June 12, 2006. Observers in New Zealand and Australia recorded the event. The results, summarized by Eliot F. Young of the Southwest Research Institute and coworkers, confirmed that Pluto's atmosphere had more than doubled in mass compared to 1988. No atmospheric collapse yet. And they ruled out the possibility, first raised in 1988, that there was a layer of haze just above Pluto's surface.

The latest word on Pluto science came in March 2009, when the European Southern Observatory announced findings from a study with its 323-inch Antu reflector of the Very Large Telescope in Chile. The researchers, led by Emmanuel Lellouch of the Paris Observatory, found that Pluto's lower atmosphere was about 40 degrees Celsius (72 degrees Fahrenheit) hotter than the surface when the observations were made in August 2008. Occultation watchers had already discovered that the upper atmosphere is even hotter. On Earth, the temperature cools with increasing altitude in the lowest layer of the atmosphere, the troposphere. In the layer above, the stratosphere, the temperature increases with height. The Lellouch findings indicate that the known regions of Pluto's atmosphere are all stratosphere. Pluto continues to surprise.

BEYOND PLUTO

In the early 1990s, astronomers began to detect other objects beyond Neptune, so Pluto was not alone. The first such Kuiper Belt object (not counting Pluto and its moons) was 1992 QB_1, which has never received an official name. By 2006, many such KBOs had been found, and so the debate over whether Pluto was a planet or just another KBO—perhaps to be dignified with the designation "dwarf planet"—intensified.

Just a month before the Hubble Space Telescope confirmed the existence of Nix and Hydra, NASA launched the New Horizons mission, led by Alan Stern. It is a fast space probe that will visit Pluto and Charon in July 2015 and go on to target at least one of the icy objects farther from the Sun in the Kuiper Belt.

Clyde Tombaugh died on January 17, 1997, just short of his 91st birthday, but the exploration of Pluto was in high gear. Whether it was a planet or not, astronomers still wanted to know as much as they could about it.

CHAPTER 9

Unveiling the Kuiper Belt

Heaven's utmost deep
Gives up her stars, and like a flock of sheep
They pass before his eye, are numbered, and roll on!

—Percy Bysshe Shelley, *Prometheus Unbound*, 1820

THE SOLAR SYSTEM'S UTMOST DEEP

To know what Pluto is, we need to understand its setting in the outer solar system—what is now called the Kuiper Belt.

On September 15, 1992, Pluto authority Alan Stern sent one of us (Maran) an e-mail message with the breathless subject line: "!!Seen THIS??" The body of the message consisted of International Astronomical Union Circular No. 5611, dated September 14 and issued by Brian Marsden at the IAU's Central Bureau for Astronomical Telegrams in Cambridge, Massachusetts. The Circular announced that astronomers David Jewitt and Jane Luu "report the discovery of a very faint object with very slow . . . retrograde near-opposition motion," detected with the University of Hawaii's 88-inch telescope on Mauna Kea. Marsden noted that his own computations indicated that the object, designated as 1992 QB$_1$, "is currently between 37 and 59 AU from the earth" and that some possible solutions for its still unknown orbit "are compatible with membership in the supposed Kuiper Belt."

Stern was so excited because many astronomers, himself included, had long suspected that there were icy bodies orbiting the Sun beyond Neptune, and now one had been found. According to Stern and his colleagues, these bodies, which we now call Kuiper Belt objects, were planetesimals—that is, objects that formed from the birth cloud of the solar system, some of which may have grown to the size of very small planets, like Pluto. Other astronomers disagreed. They, too, thought there might be many icy bodies in the outer solar system, but aside from Pluto, the Kuiper Belt might consist simply of huge numbers of comets. (A few even thought that Pluto itself might be a giant comet.)

The *New York Times* story on the find, headlined "Red Object Sighted Beyond Pluto May Be Part of Minor Planet Belt," quoted Jewitt as saying, "We think this is the first of a large number of similar objects waiting to be discovered in the outer solar system." Jewitt told Kathy Sawyer of the *Washington Post* that "[b]y studying this and similar objects we can learn about the way the planets formed."

The Kuiper Belt is named for Gerard Kuiper, the same astronomer who attempted to measure the diameter of Pluto with the 200-inch telescope in 1950. Kuiper, at the time, was almost alone in the United States as a planetary researcher; he estimated that in the late 1940s, there were only 1.5 full-time specialists in the country. Just a year after his measurements of Pluto, Kuiper published a lengthy paper, "On the Origin of the Solar System," in which he described a comprehensive set of physical phenomena that might account for the origins of planets, moons, and comets and for many of their properties. In retrospect, some of his ideas were better than others, but all were grounded in the best information and theory then available. He wrote the paper long before it became possible to carry out elaborate numerical modeling on computers, a crucial tool for our current understanding of solar system formation. Nowadays, Kuiper's 1951 article is remembered most for just two short pages in which he proposed what he called "the belt just outside proto-Neptune, i.e., 38 to 50 AU" from the Sun.

There's a controversy over what to call the Kuiper Belt. Some astronomers insist that primary credit should go to the Irishman Kenneth E. Edgeworth, a mustachioed veteran of the British Army, who served in World War I and later worked in the Sudan (a caricature preserved in a museum archive in the United Kingdom shows him sporting a fez). Two years before Kuiper's 1951 paper, Edgeworth, a scientific amateur, published his own concept for the origin of the solar system, in which he discussed the possible existence of a belt of many comets beyond Neptune and Pluto. Shortly after the discovery of 1992 QB_1, Jewitt wrote to us that "Kuiper's 1951 paper was beaten to the punch by Edgeworth's 1949 . . . paper, yet Edgy seems to have been neglected by history."

Whether because Edgeworth was an amateur or because his scientific arguments were not as convincing as Kuiper's, relatively few astronomers refer to the Edgeworth Belt, although some give credit to both theorists. It's not very surprising, given how history is written and rewritten, that the credit for the belt has gone to Kuiper. *Science* magazine called the discovery of 1992 QB_1 "a posthumous triumph" for Kuiper and compared the importance of the event with the discovery of the first asteroid in 1801. What's more surprising is that Kuiper actually rejected the idea that the belt of objects existed. He thought it began fully populated but that gravitational interactions with Pluto would have swept the belt clean of most bodies except Pluto long before the modern era. The smaller bodies might have been flung into orbits taking them in toward Jupiter, which would remove them from the belt, or they might have been flung far out into new orbits at immense distances from the Sun. A few might even have collided with Pluto, helping it grow to its present size and mass. So Kuiper concluded that his belt was virtually empty. Nowadays, some take this conclusion as grounds for keeping Kuiper's name off the belt and perhaps favoring Edgeworth. But we think that

Kuiper's reasoning made sense: he reached his erroneous conclusion because he assumed, as did other experts of the time, that Pluto was much more massive than we now know it to be. Had Kuiper known the correct mass of Pluto, he might have looked forward to the discovery of a populated belt, exactly as has transpired since Jewitt and Luu found the first KBO besides Pluto in 1992.

As a historical note, it's worth considering that if Kuiper had been right and Pluto had cleared out the belt around it of other bodies, it would have met the definition of a planet that the IAU adopted in 2006. Because the Kuiper Belt is filled with untold thousands of objects, Kuiper was wrong about the belt, and Pluto fails the IAU definition.

Kenneth Edgeworth died in 1972 at age 91, and Gerard Kuiper passed away in 1973 at 68. Both had theorized the existence of the belt beyond Neptune from their ideas about how the solar system formed, although Kuiper believed that Pluto had emptied the belt. Both men died not knowing that the belt exists. Once Charon was discovered in 1978, astronomers could reliably estimate the mass of Pluto and see that it is too small to have cleared out the belt. Then in 1979, the Uruguayan astronomer Julio A. Fernandez submitted a paper that predicted the existence of the Kuiper Belt on different grounds. He proposed that the still-hypothetical belt represented the source region of the short-period comets (as Fernandez called them). Scientists knew of 73 of these comets with orbital periods of just 3.3 to 13 years, and it was estimated that on average, each such comet would last only about 1,400 years in an orbit so close to the Sun and Jupiter. Since the solar system is over 4.5 billion years old, there must be a steady supply of new short-period comets to replace those that are destroyed. Fernandez calculated that if there are great numbers of comets in the region of the Kuiper Belt, enough of them would be perturbed by Neptune's gravity to pass inward through the solar system, where (also affected by the gravitation of Jupiter and others of the giant planets) they would enter short-period orbits. Today, astronomers call these objects the Jupiter-family comets. Fernandez calculated that some of the largest objects in the Kuiper Belt would be as bright as magnitude 17 or 18, 100 to 700 times brighter than the first KBO, 1992 QB$_1$ (magnitude 23), would prove to be, but many astronomers now accept his hypothesis that the Kuiper Belt is the source of the Jupiter-family comets.

Still another theory about the Kuiper Belt was advanced by Alan Stern in 1991, the year before the belt's existence was verified. Stern wasn't concerned with comets in the belt but with hypothetical much larger objects with diameters of perhaps 600 miles (the solid body within a comet is usually just 10 miles or so across or less). He argued that several unusual circumstances in the outer solar system could be explained if such bodies existed. The oddball situations he cited are the large tilts of the polar axes of Uranus and Neptune with respect to their orbits (98 degrees for Uranus; 30 degrees for Neptune), the presence of Neptune's large moon Triton in a retrograde orbit where it could not have been formed but must have been captured, and the presence of Pluto and Charon as a binary planet. (Retrograde means that Triton is orbiting Neptune in the direction opposite the direction Neptune orbits.) Stern pointed out that these odd circumstances could be explained by collisions or close encounters (as the case may be) with the 600-mile wide bodies that he envisioned. Uranus and Neptune might tilt over as a result; Pluto might be fractured into two parts (the smaller would become Charon, according to this theory), or Pluto might somehow capture the approaching body, which would then be Charon; and Triton might be a body from the belt, captured by Neptune. (Moons that form in place around a giant planet like Neptune revolve around the planet in a prograde manner—that is, they orbit in the same direction that the planet spins.) Since collisions of large bodies with planets are rare, there had to be many such bodies in the Kuiper Belt to make the hypothesized collisions with Uranus, Neptune, and Pluto plausible. Stern wrote, "This hypothesis implies that the present population of planets in the outer Solar System is much larger than previously recognized." By planets, he meant just what Owen Gingerich's committee later proposed for an IAU definition in 2006, which the IAU General Assembly rejected: objects orbiting the Sun that are massive enough to assume a round shape under their own gravity and that are too small to generate energy by nuclear fusion, like a star.

Julio Fernandez's theory that the Jupiter-family comets come from the Kuiper Belt, which would mean that there is a vast number of comet-sized bodies in the belt, is widely accepted, but direct evidence for the theory is controversial. While evidence for larger bodies in the belt, and thus for Alan Stern's theory of such objects, has been mounting since the discovery of 1992 QB_1, with more than a thousand now known, claimed detections of the small bodies are often unconfirmed at best. The small bodies have been sought with the Hubble Space Telescope, with a group of small automated telescopes in Taiwan, and with a large telescope on a mountain in the Canary Islands, among other projects.

The Hubble search for comets in the Kuiper Belt seemed to have borne fruit in June 1995, when Anita L. Cochran of the University of Texas, Austin, announced the discovery of what the *New York Times* called "a reservoir of icy, comet-size objects on the fringes of the solar system." Cochran was the leader of the Hubble search team, which included Alan Stern and other experts. The team reported finding about 30 of the supposed comets, each perhaps a half-dozen miles wide. But it was a statistical result, meaning that some faint images might be real and some might be just "noise," or random fluctuations in the electronic data from the Hubble camera. Cochran could not point to any one image and say that this particular one was a comet, as opposed to noise. Her fellow team member Harold F. Levison of the Southwest Research Institute asserted that the seeming discovery made the Kuiper Belt "the most populous region in the planetary system." He didn't mean that 30 comets is a big population; he was commenting on the total number of comets in the whole belt based on the small region sampled. *Texas Monthly* described Cochran as a person who "talks about her work with the full-tilt enthusiasm that other people reserve for sports teams." With Texan pride, the magazine recounted how she was led to "redefine the solar system." Yet, some researchers began to doubt the Hubble results. By October 1997, other astronomers, led by Michael E. Brown, a Kuiper Belt expert at the California Institute of Technology in Pasadena, had reanalyzed the Hubble data and published their findings in the *Astrophysical Journal*. They claimed that the supposed faint comets were all just noise in the data. The Cochran team did a reanalysis of their own and issued a spirited rebuttal in 1998, reviving their claim of detecting the little bodies. Two teams of astronomers were arguing about objects that, if real, were shining at extremely faint magnitude of about 28 or 29, beyond the limits of most observational techniques of the time. What now was needed was not another analysis of the same Hubble data, but new and better observations that would bear on the existence of comets in the Kuiper Belt.

Another attempt to detect the controversial comets in the Kuiper Belt began in January 2005, when three 20-inch robotic telescopes in Taiwan began searching for evidence of the small bodies. (A fourth telescope was later added.) In contrast to Anita Cochran's project with the Hubble Space Telescope, the researchers from Taiwan and the United States were *not* attempting to photograph the distant bodies; they were trying to record random events when a distant comet might briefly pass in front of a star, occulting the star and causing it to fade for a moment. The comet would not be seen but would betray its presence by eclipsing a star. Taking rapid-fire digital images at up to five times per second, this experiment was optimized to detect occultations by comets about two miles in diameter. But after two years of intense scrutiny, not a single occultation event that could be ascribed to an object in the Kuiper Belt had been found. The Harvard-Smithsonian Center for Astrophysics (CfA), a major partner in the Taiwanese-American Occultation Survey (TAOS), put a good face on the negative result, noting that "when astronomers don't find what they are looking for, the defeat can provide as much information as a successful search." The CfA, whose director, Charles Alcock, was an important participant in TAOS, concluded that "[t]he outer solar system . . . appears not as crowded as some theories suggest, perhaps because small KBOs have already stuck together to form larger bodies or frequent collisions have ground down small KBOs into even smaller bits below the threshold of the survey."

It's hard to be sure of the TAOS findings because they were not independently duplicated by other astronomers using other equipment. There's an old saying among scientists: "The absence of evidence is not evidence of absence." Perhaps there are comets in the

Kuiper Belt and TAOS somehow missed them. The astronomers who accept Fernandez's theory that Jupiter-family comets come from the Kuiper Belt may find it hard to accept the TAOS findings, and those who believe the TAOS findings must have some doubts about the theory.

Perhaps the TAOS telescopes were not up to their task, and an alternate method of finding the faint Kuiper Belt comets needed to be found. One such method was to look for occultations of stars by KBOs using a bigger telescope than did the TAOS group. That's just what a team of 18 astronomers from Canada, France, Italy, Portugal, Ukraine, and the United Kingdom reported doing in a paper published in 2006. They employed an advanced instrument named ULTRACAM, described as a "high-speed triple-beam imaging photometer," mounted on the 165-inch William Herschel Telescope on La Palma in the Canary Islands. Unlike the TAOS team, which did not detect a single occultation by an object in the Kuiper Belt, the Herschel Telescope team claimed that they recorded occultations by two small objects at exceptional distances, beyond 100 AU from the Sun. That distance is much farther than the region where Fernandez believed that the Jupiter-family comets originate. The purported detections came in the analysis of almost two million brief exposures and were attributed by project leader Françoise Roques of the Paris Observatory and her colleagues to objects with diameters of about 0.4 miles. Like the TAOS report, this finding has not been independently confirmed. It is not a definitive test of Fernandez's theory.

Meanwhile, NASA astronauts installed the Advanced Camera for Surveys on the Hubble Space Telescope in March 2002. An even better camera than was available to Anita Cochran in 1995, it is the same instrument that was later used to discover Pluto's small moons Nix and Hydra. Now it was the turn of another science team to search for small bodies in the Kuiper Belt with Hubble. Led by Gary M. Bernstein of the University of Pennsylvania, the new group devoted 125 orbits of the telescope to the search, which reached almost to the 29th magnitude. In September 2004, they reported that the search had turned up just three objects in the Kuiper Belt, each between 40 and 42 AU from the Sun and each one confirmed by an additional observation. These KBOs were larger than the comets sought for in the TAOS project and had estimated diameters of 30, 20, and 17 miles, but even if they were comets, there were fewer of them than expected. The Hubble ACS observations were capable of revealing more distant objects in the belt than these three, but none were found. By 2004, astronomers had recognized what amounts to two Kuiper Belts. One, the "classical Kuiper Belt," consists of objects in a flat distribution or disk. The other, the "excited Kuiper Belt," consists of objects whose orbits are noticeably inclined to the plane of the solar system, so that these KBOs move far above and below it as they orbit the Sun. The results of the Bernstein team showed that it is unlikely (but perhaps not impossible) that the classical Kuiper Belt is the source of the Jupiter-family comets. However, it seems possible, but not at all certain, that the excited Kuiper Belt is the source.

Small comets have not been found and confirmed in the Kuiper Belt, but at least one scientist looked for even smaller stuff. Vigdor Teplitz, a physicist at Southern Methodist University (who has since joined NASA), sought evidence for dust particles in the belt in a study done in the late 1990s. By then, the interplanetary space probe Pioneer 10, which had explored Jupiter as it flew by in 1973, had spent more than a decade beyond the orbits of Neptune and Pluto. Teplitz used information from Pioneer 10 (whose scientific instruments were no longer operating) to search for dust. He calculated that a high-speed collision between a suitable particle and the propellant tank on Pioneer 10 (used occasionally to stabilize the spacecraft) would cause a penetration that ruptured the tank, causing the propellant gas to swiftly empty into space. But no such propellant loss had occurred, allowing Teplitz to calculate a likely upper limit on the amount of dust in the belt. This, like the TAOS survey result, is only a negative result. But the apparent scarcity of small comets in the Kuiper Belt may fit with an apparent paucity of dust. Dust particles in the belt might be primordial, dating to the birth of the solar system, but some experts calculate that periodic events in which the whole solar system passes through a giant molecular cloud in the Milky Way as the Sun wends its way around the center of the galaxy, would sweep clear any primordial dust. On the other hand, random collisions between comets in the belt would grind them down into ever smaller bodies and liberate plenty of new dust particles. Perhaps a lack of comets is consistent with a low dust content.

LARGE BODY FEVER

Since David Jewitt and June Luu discovered the first KBO in 1992, astronomers have vied to find other relatively large objects in the Kuiper Belt. More than 1,200 KBOs with sizes above roughly 60 miles have been found, allowing detailed analysis and correlations of physical properties with orbital parameters. This has allowed Jewitt, Luu, and other investigators to classify the KBOs into distinct groups and to gather evidence for structure in the distribution of objects in space beyond Neptune. Despite the solid scientific interest in these findings, the greatest excitement and the biggest controversies have surrounded discoveries of large KBOs. When it comes to those projects, Michael Brown, the Caltech astronomer who led criticism of Anita Cochran's Hubble Space Telescope survey, has emerged as the leading light in the field and the closest thing to a rock star among trans-Neptunian scientists. He has been profiled in the *New Yorker*, the *New York Times*, and *Discover*, and his findings regularly make news.

Brown, who searches for large KBOs with the 48-inch Samuel Oschin Telescope at Palomar Observatory (and other telescopes), has been central to the issue of whether Pluto is a planet, because he discovered KBOs that are comparable in size to Pluto, and one that is even bigger. Either the IAU needed to acknowledge one or more of Brown's finds as full-fledged planets or they had to demote Pluto (at least that's what some astronomers thought). And Michael Brown was central to another highly public controversy when he suggested that two Spanish astronomers may have attempted to claim credit for discovering an important KBO that he found and was preparing to announce.

After the landmark discovery of 1992 QB$_1$, astronomers began to find other KBOs at a regular rate. Another surprise came when a group led by David Jewitt examined an object discovered in 2000 by the 36-inch Spacewatch telescope in Arizona. This KBO, named Varuna, was bright enough for the Jewitt team to measure not just its brightness in the reflected light of the Sun, but also the strength of the infrared radiation that Varuna emits because of the heat in its surface layer. Comparing these two quantities yielded a reliable estimate of Varuna's diameter, rather than a rough guess of the size as was done for the other KBOs found since 1992. The answer, give or take 15 percent: Varuna is 560 miles wide. At that point, it was the third-largest known object in the Kuiper Belt, after Pluto and Charon. Alan Stern's theory that there are many bodies the size of small planets in the Kuiper Belt was looking better than ever.

Another noteworthy finding was that, as the Jewitt team wrote in *Nature*, Varuna's "surface is darker than Pluto's, suggesting that it is largely devoid of fresh ice, but brighter than previously assumed for KBOs." Astronomers, at last, were starting to get meaningful physical information on individual KBOs. In a follow-up study, Jewitt and Scott S. Sheppard, also affiliated with the University of Hawaii, monitored the brightness of Varuna with the university's 88-inch telescope and discovered that the object was rotating once every 6.3 hours. Analysis suggested that Varuna is elongated rather than round. Jewitt and Sheppard asserted that "Varuna may be a rotationally distorted rubble pile, with a weak internal constitution due to fracturing by past impacts." They attributed this condition to an early period in the history of the solar system when big objects were much more common in the Kuiper Belt and when collisions among bodies in the belt were relatively frequent. So physical information was beginning to elucidate ancient history.

By 2002, the pace of major KBO finding was quickening, much to the delight of planetary astronomers. One exciting new object was Quaoar, which was found in June 2002 by Brown and his collaborator, Chad Trujillo. At the end of July, the two astronomers observed Quaoar with the Hubble Space Telescope's Advanced Camera for Surveys, whose high resolution enabled them to measure the body size. They found that it is about 785 miles in diameter, significantly larger than Varuna and probably larger than Charon.

Quaoar was the largest and brightest known KBO in 2002. As such, it was a good target for spectroscopic observations, which might provide evidence for specific substances on the surface of the body, like the methane ice known to be on Pluto. Jewitt and Luu looked for ices on Quaoar in May 2004 with the National Astronomical Observatory of Japan's 320-inch Subaru Telescope located on Mauna Kea. They found the characteristic spectral signature of crystalline water ice (like ice on Earth) and weaker evidence for frozen ammonia as well. The crystalline water ice was a surprise, because objects as far from the Sun as Quaoar have surface temperatures around -370 degrees Fahrenheit. Under those conditions, water should freeze as so-called amorphous ice, with no crystal structure. Furthermore, according to Jewitt and Luu, the constant bombardment of Quaoar by cosmic rays would destroy crystalline ice in only 10 million years or so, yet the solar system formed about 4.6 billion years ago. The implication is that something happened on Quaoar within the past 10 million years to convert the surface to fresh crystalline ice. Two possibilities are (1) that another body collided with Quaoar, melting any amorphous surface ice, which then froze in crystalline form before the temperature dropped back to -370 degrees Fahrenheit, and (2) that there is cryovolcanism on this KBO. We think the latter idea is more likely. Cryovolcanism means volcanic-like eruptions, not of hot lava, but of a substance that is icy and yet still much hotter than the surface of the object where the eruptions are underway. The "hotter" material would still be extremely cold compared to conditions on Earth. According to David J. Stevenson, an expert on planetary interiors at Caltech, the rock that probably exists inside Quaoar would contain radioactive isotopes that generate heat. The heat would flow up into the ice layer above, where it might stimulate cryovolcanism that would spew "lava" that would then freeze in the form of crystalline ice. Another possibility that Stevenson noted in *Nature* in 2004 is that there might be as-yet-unidentified physical processes on the Quaoar surface that convert amorphous water ice to the crystalline form without the application of much heat. KBOs were getting, as Alice said in Wonderland, "curiouser and curiouser."

Another major announcement in 2002 was the discovery that a KBO found earlier, 1998 WW$_{31}$, is actually a binary system consisting of two objects that are each about 60 miles in diameter. 1998 WW$_{31}$ was observed with the 142-inch Canada-France-Hawaii Telescope (CFHT), also on Mauna Kea, on two nights in December 2000, but the object was not readily apparent in the images until CFHT director Christian Veillet reprocessed the data in April 2001. He not only saw WW$_{31}$ in the reprocessed observations, but saw that there were two bodies present in close proximity. If you don't count Pluto-Charon, this was the first known binary Kuiper Belt object. Since 2000, a few dozen binary KBOs have been discovered, mostly with the Hubble Space Telescope, and there probably are many more, including some systems in which the two objects are so close that they may be almost touching.

Some binary KBOs, like Pluto and Charon (actually a quadruple KBO when you count the small moons, Nix and Hydra), may have been created by collisions that broke a preexisting object into two (or more) parts. But Keith Noll, an astronomer at the Space Telescope Science Institute, finds that the collision theory is unlikely to apply in most cases, given that most of the binary KBOs have nearly equal-sized members, an unlikely result of collisions. David Jewitt told us recently that the origin of the binary systems remains to be proven.

The year 2003 was a milestone in the exploration of the Kuiper Belt, with the discovery of three large objects by Michael Brown and his collaborators: Eris, which brought the widespread argument about the nature of Pluto to a head; Haumea, the KBO whose discovery credit, by some accounts, almost was stolen; and Sedna, located far beyond the Kuiper Belt in an orbit that Brown, Trujillo, and coworker David Rabinowitz of Yale University described as "unexpected in our current understanding of the solar system."

Eris, first known as 2003 UB$_{313}$, and not publicly announced until 2005, was later named for the Greek goddess of warfare and strife. It was the first object found beyond Neptune that is larger than Pluto in diameter and almost 30 percent greater in mass as well. Now that objects more massive than Pluto were showing up in the outer solar system, the International Astronomical Union had to decide whether to classify them as planets or as another class of object for naming purposes, as we described in Chapter 2. Noting the goddess Eris's traditional association with strife, Brown remarked on his Web site that the physical body Eris "stirred up a great deal of trouble among the international astronomical community when the question of its proper designation led to a raucous meeting of the IAU in Prague." There was little doubt that Eris actually was more massive than Pluto, because on September 10, 2005, 11 months before the Prague meeting, Brown and coworkers used one of the two 400-inch Keck telescopes on Mauna Kea, equipped with a laser-assisted optical system that eliminates nearly all of the blurring caused by air turbulence above the telescope, to make high-resolution images of the object. The new images revealed that Eris has a moon, later named Dysnomia, which revolves around Eris about once every 16 days. The orbit of Dysnomia, established with further images from the Hubble Space Telescope, allowed Brown to calculate the mass of Eris, just as the orbit of Charon allowed Robert Harrington to estimate the mass of Pluto in 1978. Brown stated that "Eris has a mass 27% higher that that of Pluto (with an uncertainty of only 2%)."

Haumea was first designated as 2003 EL_{61} by the IAU Minor Planet Center, apparently because the earliest photograph of the object by Brown's discovery team dated to 2003. However, Govert Schilling in *The Hunt for Planet X* writes that Haumea (as the IAU later named it) was not actually recognized in the team's data until December 28, 2004. It was clear that this was a very unusual object, and Brown planned to gather more data and to announce and describe 2003 EL_{61} at a conference in September 2005.

Here's why Haumea is special: it is spinning rapidly (making one turn around its axis in less than four hours); it is elongated like a cigar or a long, narrow oval; and it is so dense that it must be mostly rock, with perhaps a relatively thin outer layer of ice. Along the length of the "cigar," the object is about Pluto's size, but at right angles to that axis, it's much narrower, and it is a few times less massive than Pluto. The remarkable object also has at least two moons.

Brown submitted brief abstracts of papers that he intended to present at the September conference. The abstracts, now out of his hands, went online in July 2005, according to the "electronic trail" that Brown prepared of the ensuing controversy. In three abstracts, he used an object name for Haumea that was just an internal reference number used within his own team, which should convey no information on the position of the body, for example. The reference number was K405060A, and the abstracts mentioned that this object might be larger and brighter than any previously known KBO. A remark like that was bound to generate interest among other scientists.

Just one week after the abstracts went online, according to the Brown timeline, the IAU Minor Planet Center, the clearinghouse for asteroid and KBO discoveries, received e-mail from a Spanish astronomer reporting the discovery made with a colleague of a new object in observations that had been made in 2003. It was clear at once to Brown that this was the same object as Haumea.

Brown at first thought he had simply been beaten to the discovery by the two Spanish astronomers. But then he learned that observing records of the 51-inch telescope at Cerro Tololo Inter-American Observatory in Chile, which was used in his observations of Haumea, had been remotely examined with a computer in Spain—the same computer that sent the "discovery" e-mail message to the Minor Planet Center. It turned out, by Brown's account, that a Google search on the meaningless internal reference number K405060A contained in his abstracts could lead a computer user to the records of Brown's observations with the telescope in Chile, which included that reference number and the positions of the object on the nights of observation.

Soon Brown was crying foul, and on September 13, as the controversy escalated, the *New York Times* printed a story headlined "One Find, Two Astronomers: An Ethical Brawl." The Spanish astronomers were identified in the story as Jose Luis Ortiz and student Pablo Santos-Sanz. According to Dennis Overbye of the *Times*, Brown wrote to the Minor Planet Center on August 15, 2005, requesting that the Spaniards "be stripped of the official discovery status" and that the IAU also "issue a statement condemning their actions." The IAU doesn't have formal procedures for adjudicating a dispute like this. The Spanish astronomers have denied any wrong-doing and reportedly have criticized Brown for not having released his discovery information promptly.

The controversy over Haumea has not been officially resolved. Rather, like several past disagreements in planetary astronomy that we recount in earlier chapters, the general sentiment of astronomers has simply converged on an informal de facto judgment. Discovery credit is generally given to the Brown team, but as time goes on, we believe researchers will concentrate on the scientific import of Haumea rather than on this unseemly chapter in its history.

Then there's Sedna, the last of the three brilliant 2003 discoveries. It's truly extraordinary, orbiting the Sun far outside the Kuiper Belt in a highly elongated orbit that takes it from perihelion at 76 AU, almost double Pluto's average distance from the Sun, to aphelion at an amazing 975 AU, so far out that it takes 120 centuries (versus 248 years for Pluto) to make one revolution around the Sun. Sedna may be roughly as large as Haumea. It is believed that a huge swarm of comets, ejected long ago from the Kuiper Belt or other regions "close" to the Sun, exist at vast distances in a great swarm called the Oort Cloud after the Dutch astronomer Jan Oort, who first imagined it. Brown's group has referred to Sedna as a "candidate inner Oort cloud planetoid." Whatever Sedna may be, its discovery shows that searches through the Kuiper Belt, despite their successes, are just a promising start to the exploration of the far reaches of the solar system. The regions beyond the Kuiper Belt may be every bit as remarkable as the parts of the solar system that we have explored so far.

CHAPTER 10

Exoplanet Wars

I believe the Planets are Worlds about the Sun, and that the fixed Stars are also Suns, which have Planets about them, that's to say, Worlds, which because of their smallness, and that their borrowed light cannot reach us, are not discernible by Men in this World.

—Cyrano de Bergerac, *The Other World: The Comical History of the States and Empires of the Moon*, 1657

NEW WORLDS BEYOND

For centuries, scientists and philosophers wondered if there were other worlds beyond our solar system, and if there were, did they harbor life? Today we know of about 40 times as many planets of stars other than the Sun (called exoplanets, or extrasolar planets) as there are planets in our solar system (whether you count Pluto as a planet or not). There is no evidence of life on these planets, but since our study of extrasolar planets is just in its infancy, there is still insufficient data to draw any firm conclusions about the presence or absence of life outside our solar system.

Before these discoveries, there were many astronomers who believed that the solar system was unique among the hundreds of billions of stars in our galaxy, or that there were at most only a handful of systems like ours. But that opinion was based on an incomplete understanding of the nature of planetary systems. We didn't know how planets form, and when scientists don't comprehend how something happens in nature, they often think that the unknown process must involve an extremely unlikely event. Since the mid-1990s, we have learned that exoplanets are common, and the consensus of astronomers is that planets come about as a natural consequence of star formation, rather than as an extraordinary phenomenon. The era of the first exoplanet discoveries occurred about the same time that astronomers using the Hubble Space Telescope and other telescopes were finding round, flattened clouds around newborn stars. These are the clouds in which physical conditions may be conducive to planet formation.

Not one of the more than 330 known exoplanets satisfies the IAU definition for a planet, because (like Pluto) they are all specifically excluded. The IAU doesn't call them dwarf planets, as it refers to Pluto, and it doesn't say that they are something other than true planets. The IAU simply limits its planetary classification system to objects in the solar system. What constitutes an exoplanet or evidence of an exoplanet is left to the judgment of individual astronomers. And they often disagree.

According to Brian Marsden, the initiator of the campaign to demote Pluto, "If an astronomer wishes to make front-page news, the surest way for him to proceed is to claim the discovery of a major planet." He apparently had the solar system in mind, but Marsden's remark applies equally well to exoplanets. During the two decades after 1963 and especially in the decade after 1984, there was one controversial announcement after another that the first exoplanet had been discovered or that "the best evidence yet" for exoplanet existence had been found. Almost without exception, these claims were criticized and rejected by skeptical colleagues in the astronomical community or even withdrawn by the scientists who made them. (Stephen Maran presided at several press conferences where these "discoveries" were announced and also chaired one news briefing where the claim of a "planet" was retracted by the same astronomer who had recently reported finding it.)

Complicating the situation was the fact that while some astronomers were looking for still-hypothetical exoplanets, others were looking for another class of hypothetical objects called brown dwarfs. Brown dwarfs (which have since been found) are objects that probably form by condensing from clouds of interstellar gas and dust, just like stars, but they are less massive than "true" stars. Specifically, they are not massive enough to sustain the nuclear burning of hydrogen in their central regions for long periods of time, so they don't shine brightly, and they cool down and fade away. There is no official definition of any of these terms—brown dwarf, star, or exoplanet. But many astronomers use the following working definitions: brown dwarfs have masses in the range of about 13 to 75 times the mass of Jupiter, stars have masses of more than 75 Jupiters, and exoplanets are objects born in flattened clouds ("discs") around stars and have masses below that of 13 Jupiters. True exoplanets presumably also must have more than a certain minimum mass that is not usually stated, since there are also claims that smaller objects— asteroids and comets—exist around certain stars. Further, as new results come in, individual astronomers or teams of astronomers rewrite some of these rules in the light of their own findings. As recently as 2009, some newfound objects with masses estimated at less than 10 Jupiters, and which are not orbiting stars, were termed brown dwarfs. Perhaps there's not one rule that fits all exoplanets and brown dwarfs.

PREMATURE OPTIMISM

The first widely discussed claim of an exoplanet came on April 19, 1963, when the *New York Times* trumpeted "Another Solar System Is Found 36 Trillion Miles from the Sun." The unsigned article, reporting on a meeting of the American Astronomical Society in Tucson,

Arizona, highlighted the work of Peter van de Kamp of Swarthmore College in Pennsylvania, who had announced the finding of a Jupiter-sized planet orbiting one of the closest stars to the Sun. Barnard's star, as the star is known, is a red dwarf, much cooler and fainter than the Sun. Although not visible to the naked eye, it was already famous to astronomers as the star with the highest known proper motion. In other words, as seen from Earth, Barnard's star moves faster across the background of distant stars than any other star.

Van de Kamp, the director of Swarthmore's Sproul Observatory, had measured the proper motion of Barnard's star on thousands of photographic plates made over almost 50 years with the observatory's 24-inch telescope. He found that as the star moved across the constellation Ophiuchus, a motion that corresponded to a shift in position of just over a half-millimeter per year on the photographic plates, it was also slowly wobbling. Van de Kamp interpreted the wobble as the motion of Barnard's star around a common center of mass with an unseen companion with a period of 24 years. The companion, whose mass appeared to be 1.5 times that of Jupiter, was supposedly an exoplanet.

What news reports did not mention was that the wobble was minuscule, at the utmost limit of measurability. On van de Kamp's photographs, it amounted to a back-and-forth motion of only one-hundredth the size of the star's image. Furthermore, the star image itself was only one-ninth of a millimeter wide. (There are 25.4 millimeters in an inch.) Given the technology of the day, that was an extremely small wobble to detect, and soon other experts challenged whether it was real. Chief among them were two astronomers, Robert Harrington and George Gatewood. Harrington, a former student of van de Kamp and the same person who first calculated the orbit of Pluto's moon Charon, worked at the U.S. Naval Observatory, and Gatewood was a professor who studied the motions of stars at the University of Pittsburgh's Allegheny Observatory.

Both Harrington and Gatewood had separate collections of telescopic photographs of Barnard's star, and they found no credible wobble in their measurements of these plates. But neither they nor any other critic had plates of the star taken over as many years as did van de Kamp. And none of them was as eminent as he in the field of astrometry, the precision measurement of the positions of stars. In the introduction to van de Kamp's 1986 work, *Dark Companions of Stars*, W. Butler Burton, a respected astronomer at the Leiden Observatory in the Netherlands, called van de Kamp "the master" of astrometry, which, Burton wrote, "requires that its practitioners be meticulous, patient and persistent." Perhaps the naysayers were not as meticulous and patient as van de Kamp; certainly he had persisted at the study of Barnard's star much longer than they.

By the late 1960s, van de Kamp had refined his analysis, first revising the orbital period of the Barnard star planet from 24 to 25 years and then reinterpreting his measurements, finding that they might be best explained by two planets, each less massive than Jupiter (0.8 and 0.4 Jupiter masses), one orbiting in 26 years and one in 12 years. If you believed in van de Kamp's work, this refinement meant that with additional data and analysis, he had been able to discern more of the Barnard's star system. But if you were one of the doubters, you might suppose that the new data were not consistent with the wobble motion that was first inferred, and that lacking a smooth 24- or 25-year wobble in the total set of old and new data, van de Kamp was grasping at straws—fitting data with the complex superposition of the effects of two orbiting planets when, in fact, the data were not good enough to substantiate that any planet was present.

Despite repeated observations, neither Harrington nor Gatewood ever detected a wobble in Barnard's star. Peter van de Kamp died in 1995 at age 93, probably still believing that his discovery was valid. By 1973, another astronomer had discovered that measurements of stars with the 24-inch Sproul telescope showed inconsistencies with earlier measurements after 1949 and again after 1957, two years in which adjustments were made in the way in which the main lens was mounted on the telescope. Van de Kamp dismissed the argument that these adjustments could produce spurious motions in his analysis of Barnard's star, but many astronomers now think that this was the case. In 1999, observers led by G. Fritz Benedict of the University of Texas reported that when they studied Barnard's star with the Hubble Space Telescope, they could find no hint of a planet-induced wobble to a precision substantially exceeding that which van de Kamp was capable of.

Van de Kamp once quoted the Talmud: "If you want to understand the invisible, look careful at the visible." Perhaps he was not careful enough. But if so, neither were most of the other astronomers who made controversial claims of exoplanet discoveries in the years to come.

ANOTHER UNSTEADY STAR

The story of Barnard's star was fairly straightforward compared to that of VB 8B, which burst onto the astronomical scene in 1984. Barnard's star pitted one astronomer's claim of extrasolar planets against the skepticism of much of the astronomical community. In the case of VB 8B (a companion of the faint red dwarf star Van Biesbroeck 8 (VB 8A)), the battle began with dueling press releases by two respected research institutions, each of which claimed credit for the discovery. The National Science Foundation's December 11, 1984, announcement said that VB 8B was "what may be the first planet ever observed outside the solar system" and that "[i]f the discovery is verified, it would climax a centuries-old quest to find such a body."

The discovery team referred to by the NSF was led by Donald W. McCarthy of the University of Arizona. The *New York Times*, which had been tipped off to the impending announcement, ran a story on the front page on the morning of December 11, headlined "Possible Planet Found Beyond Solar System." On the other hand, the U.S. Naval Observatory's press release, also dated December 11 but clearly issued in response to the NSF claim and probably triggered by the *Times* story, asserted that Robert Harrington and his coworkers had discovered VB 8B first and had already published the find in the *Astronomical Journal*. Furthermore, that press release, titled "USNO ASTRONOMERS SAY NEW OBJECT IS NOT A PLANET," reported that VB 8B "is a Brown Dwarf and NOT a planetary object," an assertion credited to Harrington. It seemed that either the first exoplanet or the first brown dwarf had been found. Either way, it would be a landmark discovery.

McCarthy supposedly had obtained an image of VB 8B with an advanced new technique (familiar to very few astronomers in 1984) called infrared speckle interferometry. Harrington had no image of VB 8B. His team had used astrometry, the same method that Peter van de Kamp employed in studying Barnard's star: they had found a wobble in the motion of VB 8, indicating the presence of a tiny, unseen companion.

About a month after the press releases came out, Harrington and McCarthy described their respective observations in a press conference at an American Astronomical Society meeting. By an odd coincidence, the meeting was in Tucson, where van de Kamp had made his announcement on Barnard's star almost 21 years before. The two astronomers treated each other cordially, and it was clear that both now agreed on "brown dwarf" as the most likely description of VB 8B. The first known brown dwarf had been found—or so it seemed. But by 1987, there was an astonishing reversal. Two other teams, using telescopes respectively in Chile and Hawaii, looked for VB 8B and said it wasn't there. And when both McCarthy and Harrington looked again, each using his own technique, neither could reproduce the previous detection. VB 8B was an astounding case of an astronomical object independently "discovered" by two groups, each of which was mistaken. *Sky & Telescope* magazine speculated that the errors may have come from "subtle systematic effects of Earth's atmosphere on the observations." There was still no confirmed case of an exoplanet or a brown dwarf. The searchers were back to square one.

CAMPBELL'S PLANETS

The next batter to step up the plate at the World Series of exoplanet discovery announcements was Bruce Campbell, a bright young Canadian astronomer at the Dominion Astrophysical Observatory near Victoria, British Columbia. Campbell thrilled a jam-packed press conference in Vancouver, BC, on June 18, 1987, with his purported discovery of a veritable soup of planets. Campbell had been examining 16 stars for over 6 years with the 142-inch Canada-France-Hawaii Telescope using a new method for making extremely precise measurements of stellar spectra. Remarkably, 7 of these 16 stars showed evidence of possible planets, ranging in estimated mass from 1 to 10 times the mass of Jupiter. If all 7 planets were real, it might indicate that extrasolar planets were extremely common; assuming that Campbell's sample was typical, maybe half of stars like our Sun might have planets.

As Campbell spoke, astronomers crowded into the press room, sitting in rows behind the smaller group of reporters. It was unusual to have so many astronomers attending an event usually attended by science writers, but the announcement commanded interest and respect because Campbell's senior and principal collaborator, Gordon A. H. Walker of the University of British Columbia (UBC) in Vancouver, was a highly regarded expert on the type of measurements central to the claimed discovery. Instead of measuring wobbles in the motions of stars across the sky as van de Kamp had attempted, they sought wobbles in the motions of stars along the line of sight toward and away from Earth. (These motions are called radial velocities.)

According to a well-known principle of physics, the Doppler effect, the spectrum of a star shifts toward longer wavelengths (red shift) when the star is moving away from Earth and toward shorter wavelengths (blue shift) when it is approaching the Earth. This principle guided decades of study of binary stars, allowing astronomers to derive the physical properties of pairs of stars in orbit around each other. The Campbell team employed a similar method, looking for the minute back-and-forth wobble in the radial velocity of a star that indicated the presence of an orbiting object much less massive than a binary companion—a planet, in other words.

The search for such tiny Doppler shifts required more precise spectroscopic measurements than those employed in looking for binary stars. Campbell, Walker, and coauthor Stephenson Yang of UBC obtained this high precision by using a new technique in which starlight from their telescope was passed through a sealed transparent container, or cell, that was filled with the gas hydrogen fluoride (HF). HF is a dangerous substance requiring great care in its use; in the presence of any moisture—like in the inside of one's nasal passages—it becomes a toxic and highly corrosive acid. However, the effort was deemed worthy of the risk, because HF absorbs light at many known wavelengths, producing a grid of narrow dark lines in the telescopic spectrum of a star. This reference grid allowed the Canadian astronomers to search for the extremely small Doppler shifts in the spectrum of a star caused by its orbit around the center of mass of the star and a planet.

Bruce Campbell's announcement made headlines. The *Sacramento Bee* quoted him on the first page: "I think this is the best evidence of other planetary systems to date." Astronomers and journalists alike stressed that the planet detections needed to be confirmed. But whether the findings themselves were correct or not, the Associated Press quoted the director of planetary research at Lowell Observatory as stating that the method Campbell and Walker devised to measure the wobbles "appears to be a major step forward." In California, astronomers Geoffrey Marcy and Paul Butler were already developing a modification of the Canadian method that could be used to repeat Campbell's observations or search for other planets with even higher accuracy.

Campbell's findings also excited popular interest, because one of his best cases for an exoplanet was a supposed companion of the star Epsilon Eridani. This star is visible to the naked eye, is relatively close to the Earth, and was one of the two stellar targets of the first attempt to detect radio communications from hypothetical advanced civilizations that might exist on planets of stars beyond the Sun. That experiment, called Project Ozma, was conducted by Frank Drake at the National Radio Astronomy Observatory in Green Bank, West Virginia, in 1960. Drake heard no alien broadcast, but science fiction authors and scriptwriters worked Epsilon Eridani into their stories and television programs for years thereafter. Some fans of *Star Trek* even claim that Vulcan, the fictional home planet of Mr. Spock, revolved around Epsilon Eridani, although apparently more "Trekkies" identify the Vulcan home system with another star in

Eridanus. (Don't confuse Spock's Vulcan with the purported planet Vulcan in our own solar system, as discussed in Chapter 6.)

Unfortunately, the exoplanet findings announced by Campbell could not be reproduced by follow-up observations and were eventually dismissed by other experts. Most likely, they were not quite sensitive enough for the intended purpose. Meanwhile, Campbell left the astronomy profession, complaining that Canada did not allocate sufficient funds for young astronomers like himself to gain permanent jobs in their field.

PROBLEMATIC PULSAR PLANETS

Would exoplanets ever be found? That question seemed to be answered when the July 25, 1991, issue of *Nature* ran the words "FIRST PLANET OUTSIDE THE SOLAR SYSTEM" on the cover. This text ballyhooed a brief technical paper inside the issue. In the paper, three astronomers presented the seeming discovery, by arguably the most precise method yet, of a planet in a wholly unexpected place. The authors, Matthew Bailes, Andrew G. Lyne, and Setnam L. Shemar, concluded that "there is strong evidence" that "we have observed a planet" of about 10 times the mass of Earth or somewhat more "in a six-month circular orbit with a radius of 0.7 AU." It was a planet heavier than Earth, much lighter than Jupiter (which has 318 times the mass of Earth), in an orbit very like that of Venus (whose orbital radius is 0.72 AU).

The funny thing was that Bailes, Lyne, and Shemar were radio astronomers at the University of Manchester in the United Kingdom, who used the 240-foot dish antenna (Lovell Telescope) at the Jodrell Bank Observatory. They had not watched a star wobble across the sky like van de Kamp or Harrington, had not attempted to record an image of a planet next to a star like McCarthy, and had not measured spectroscopic wobbles in the radial velocity of a star like Campbell. They didn't even observe a sunlike star or stars as all of these other astronomers had done. They had monitored the regularly repeating brief blips of radio energy, or "pulses," received by the big radio telescope from a pulsar, a tiny, dense, invisible star, no longer shining from nuclear energy like the Sun. It was left over from the supernova explosion of a massive star. As the pulsar, known as PSR 1829-10, revolved around its common center of gravity with the planet, sometimes the tiny star was closer to Earth in that little orbit, so its pulses arrived a bit earlier than expected, because they had less distance to travel to Earth than when the pulsar was at the far point in its orbit. And of course, pulses that left the pulsar when it was at the far point took longer to reach the Earth and arrived late. The three astronomers had detected a seeming wobble in the arrival time of the pulses at the Earth. The wobble only amounted to eight milliseconds, but in pulsar timing, that's almost an eternity. Unlike van de Kamp's claimed astrometric wobble of Barnard's star or Campbell's reported Doppler wobble of Epsilon Eridani, the pulse arrival time wobble of PSR 1829-10 was not near the limit of the method of measurement; it was a piece of cake.

No one was seriously expecting pulsar planets. Would a planet survive the supernova explosion? (If so, surely any life on the planet would be wiped out.) Or was it possible that a small portion of the debris from the supernova was not flung away into space but stayed in orbit around the neutron star that remained in its aftermath? (Pulsars are neutron stars from which we can detect pulses.) Then, the orbiting debris might have accumulated into one or more new planets.

Astronomers were excited about the new find, and the senior member of the pulsar team, Andrew Lyne, was invited to address the next meeting of the American Astronomical Society in January 1992 at Atlanta, Georgia. Some experts remained skeptical, however. They worried about the coincidence that the planet's orbital period was six months, half the Earth's period around the Sun. When scientists analyze astrometric measurements or Doppler shift measurements, they have to make careful corrections for the changing position of the Earth as it moves along its orbit and similar corrections for the changing position of their telescopes as the Earth turns once a day on its axis. The same kinds of numerical corrections have to be done properly when they analyze the arrival times of pulses. Otherwise they could make a serious mistake. As the saying goes, "The Devil is in the details." David Black, a theorist of planetary systems in Texas, described possible origins of the pulsar planet, at the same time cautioning that the planet might not exist: "The history of the searches for other planetary systems is littered with published detections that vanish under further scrutiny."

A week or two before Andrew Lyne set out for the January meeting in Atlanta where he would lecture on the planet discovery, he sat down to double-check his analysis of the data on PSR 1829-10. He soon realized that there had been a critical error: his computer program simulated the Earth's orbit as a perfect circle, though it is actually slightly elliptical. This error had led the analysis to identify a six-month periodic wobble in the arrival times of the pulses, when in fact there was none. The planet discovery was wrong. As described by Geoff McNamara in *Clocks in the Sky: The Story of Pulsars*, Lyne "just sat there frozen to his chair for the next half hour as the enormity of the mistake sunk in."

Lyne could have canceled his trip to Atlanta and avoided public embarrassment. According to McNamara, "He could have simply published a retraction, but in a display of extraordinary scientific courage and honesty Lyne decided to announce the error" at the meeting. Astronomers have known that the orbits of the Earth and other planets are ellipses since Kepler figured it out centuries ago, so you might guess that Lyne's confession would have drawn some knowing glances and even a snicker or two among the audience. In fact, it was greeted with loud applause from the large crowd of astronomers, who respected him for his frank admission of error. The California planet hunter Geoffrey Marcy later wrote that Lyne's "honest retraction highlighted the scientific process at its finest." Yet, although Lyne's planet wasn't real, exoplanets were hypothetical no longer. Another radio astronomy team had just discovered pulsar planets, and this time the claim was for real.

AT LAST, EXOPLANETS

At the same January 1992 meeting where Lyne confessed his error, Alexander Wolszczan of the Arecibo Observatory (AO) staff and Dale

A. Frail of the National Radio Astronomy Observatory (NRAO) in Socorro, New Mexico, announced the discovery of two companions of the pulsar PSR B1257+12, detected from wobbles in pulse arrival times with the 1,000-foot radio telescope at the AO. The planets had masses of 2.8 and 3.4 Earth masses and were located 0.47 and 0.36 AU from the pulsar, with orbital periods of about 98 and 67 Earth days, respectively.

The two pulsar planets were soon verified by another team of astronomers, who used the NRAO's 140-foot radio telescope at Green Bank, West Virginia, and before long, a third and smaller planet with roughly the mass of our Moon was discovered. Wolszczan, who had joined Pennsylvania State University in University Park by 1994, wrote at the time that the detection of the three companions of PSR B1257+12 "constitutes irrefutable evidence that the first planetary system around a star other than the sun has been identified."

Yet these first planets are often treated as almost a footnote in the story of exoplanets. That's unfair to Wolszczan and Frail, but astronomers are overwhelmingly concerned with the planets of sun-like stars, which presumably formed in the same way as the planets of the Sun and which in some cases might bear liquid water like the Earth, and even life. The pulsar planets, on the other hand, are orbiting neutron stars, the dead cinders of suns. There could be no forests, oceans, or sunny grasslands on planets such as these, raked by intense radiation from the nearby pulsar. Though the pulsar planets were undoubtedly the first extra solar system to be discovered, the search for planets around sunlike stars went on with renewed energy.

A PLETHORA OF PLANETS

In the public relations business, you announce your government's or corporation's bad news on Friday afternoon, when it is least likely to raise a fuss because fewer people read the newspaper on Saturday than any other day of the week. But the exoplanet breakthrough that astronomers were waiting for came on Friday, October 6, 1995. Two Swiss astronomers, Michael Mayor and his graduate student Didier

Queloz, told a conference in Florence, Italy, that they had found a planet of the star 51 Pegasi. Scientists from 37 countries were at the meeting, and the word spread around the world immediately. Some of us remember exactly where we were when we heard about it. 51 Peg, as it is known, is dimly visible to the naked eye, just outside the Great Square of Pegasus, a familiar star pattern in the autumn sky (for those who live in the Northern Hemisphere). Amazingly, the planet, 51 Peg B, although comparable in mass to Jupiter, was in a tiny orbit, taking it once around its sun—a planetary year—in just four Earth days. The planet hurtles around the orbit at almost 300,000 miles per hour.

The *Washington Post* ran the story of the Swiss discovery on Sunday, when the best editors are off duty. The slightly erroneous headline stated that "Italian Astronomers Say a Jupiter-like Planet Circles a Star in Pegasus." The *Post* is traditionally much less interested in pure science than its chief rival, the *New York Times*. The story ran on page 36 and noted that "[t]he announcement was greeted by heavy applause and some doubt." One criticism was ascribed to Italy's leading astronomer, Franco Pacini, who reportedly asked, "How long could a planet last so close to the principal star without evaporating from the effect of the enormous quantity of energy it absorbed?" In the solar system, the smallest planetary orbit belongs to Mercury, but it's much larger than the orbit of 51 Peg B. Mercury easily withstands the blistering heat of the Sun, because it is a world composed of iron and rock. Jupiter, like all gas giant planets in the solar system, is more than 480 million miles from the Sun. Jupiter, the most massive planet in the solar system, takes almost 12 Earth years to make 1 trip around the Sun. Theories of what exoplanetary systems would look like, published before any exoplanet was found, were based on computer simulations. The theorists who used the computers reported similar results: little rocky planets would be close to their parent stars, and gas giants would be far. In other words, the results just happened to resemble the only planetary system known before the computations were made—our own.

The Swiss discovery was made by the Doppler-wobble method with an advanced new spectrograph on the 75-inch telescope at

Haute-Provence Observatory in France. Alerted by the news, another team of planet hunters at San Francisco State University in California quickly checked on 51 Peg B with the 120-inch telescope at Lick Observatory. The team, Geoffrey Marcy and Paul Butler, had developed the world's most powerful Doppler equipment for searching out planets. Marcy and Butler used a modified version of Bruce Campbell's technique in which they replaced the hydrogen fluoride cell with one filled with iodine vapor. It allowed them to measure Doppler shifts that are four times smaller than the Swiss could observe.

Not only did Marcy and Butler have the most sensitive spectrograph for detecting planets, but they were already well into a program of monitoring sunlike stars for telltale wobbles, and so it is ironic that they had missed out on the priority of discovery. But alas, they had deliberately excluded 51 Peg from their target list. Through one of those quirks of fate (or sometimes strokes of luck) that often decide which people go down in history for accomplishing a "first," the entry for 51 Peg in a star catalog that the Californians used to select their targets was erroneous. It said that 51 Peg was a subgiant star, different from the Sun in that it was no longer burning hydrogen in its core, which had turned to helium, but was burning hydrogen in a layer above the core. This would have the effect of making the star much bigger and brighter than a "normal" star like the Sun. Marcy and Butler were concentrating on normal stars. But the catalog was wrong; 51 Peg is not a subgiant but a normal star.

Although Marcy and Butler rapidly and easily verified the 51 Peg planet, other astronomers continued in attempts to poke holes in the Swiss discovery and even in the Marcy and Butler confirmation. Some thought that the star 51 Peg was pulsating every four days, which would cause Doppler shifts that could be misinterpreted as a wobble because of orbital motion. Others pointed to magnetism and starspots (like huge versions of sunspots on our own star) as effects that might cause errors in interpreting 51 Peg's spectrum. And of course theorists, still under the spell of their earlier computations, which predicted that close-in exoplanets would be small and rocky, questioned how a gas giant planet could form so close to a star. One theorist, David Black, director of the Lunar and Planetary Institute in Houston, wrote in *Sky & Telescope*, "The jury is still out, but at this time I would bet on 51 Pegasi B being a brown dwarf, not a member of a planetary system." A storm of controversy was gathering.

Although Marcy and Butler went on in the months after the Swiss announcement to discover other exoplanets by the Doppler-wobble method, and although experts at other observatories added their own independent confirmations of the four-day wobble in the spectrum of 51 Peg, a huge challenge to the very basis of these discoveries arose in February 1997. The widely respected astronomer David F. Gray of the University of Western Ontario stated that the four-day wobble was not an orbital effect. Gray, a world authority on the atmospheres of stars and on the interpretation of star spectra, is the author of *The Observation and Analysis of Stellar Photospheres*, first published in 1976 (and now in its third edition). To scientists such as Marcy and Butler, Gray's book was like a bible in their field. But as Gray explained in *Nature*, he analyzed his own observations of 51 Peg and found that the detailed shapes of certain features in its spectrum, called absorption lines, were varying with a period and by an amount that were comparable to the claimed orbital Doppler wobbles. Gray wrote, "As the presence of a planet will not influence the shapes of spectral lines, these variations are likely to reflect a hitherto unknown mode of stellar oscillation. The presence of a planet is not required to explain the data."

The exoplanetary field was once again engulfed in controversy, and the press, which had hailed the series of exoplanet discoveries that began with 51 Peg B, was now consumed with doubt. *Scientific American* asked, "Could the first planet discovered around a sunlike star be a mirage?" *Science* titled their story "Is First Extrasolar Planet a Lost World?" The *Dallas Morning News* quoted Gray as telling a conference in Massachusetts, "I've been amazed at the excitement, publicity and emotion this issue has raised." The story was headlined "Debate rages over existence of far-off planet."

The Swiss and San Francisco groups posted strong arguments on their Web pages to rebut Gray, who defended his criticism of the 51 Peg planet on the Internet as well. But in the end, general acceptance of that exoplanet (and by extension other likely planets found by the

Doppler-wobble method) came through further work (and a frank admission) by Gray himself. In January 1998, Gray published new findings in *Nature*. He had made fresh observations of 51 Peg that did not show the suspicious changes in spectral line shapes that had triggered his rejection of the planet finding. He acknowledged that the observations used in his earlier, critical paper were too widely spaced in time to be a reliable check on a four-day periodic effect. Importantly, he also cited new measurements on 51 Peg by other groups, who also found no changes in the shapes of absorption lines, and he conceded that "a planet may indeed be the best explanation" for the Doppler wobbles found by Mayor and Queloz and by Marcy and Butler. Marcy, who by then had moved across San Francisco Bay to the University of California at Berkeley, graciously wrote that those astronomers like himself who were upset with Gray's first paper "may occasionally forget that competition and human emotion have always provided fuel for the vigorous pursuit of alternative theories. It was right that the planet interpretation should not go unchallenged."

EXOPLANETS IN THE PROMISED LAND

Since the discovery of 51 Peg B, astronomers have gone on to discover more than 330 exoplanets, mostly by the Doppler-wobble method, which experts call radial velocity monitoring. A major new way of discovering exoplanets has emerged: transit monitoring, in which observers detect the presence of planets through a tiny decrease in the brightness of a star as a planet passes in front of it. Transit observers have detected and identified gases in the atmospheres of such "transit planets" when they were close to the edge of their suns, much as Pluto watchers discovered the atmosphere of that planet when it passed in front of a star in June 1988. The European Space Agency has launched the COROT satellite to discover more transit planets, and in 2009, NASA launched Kepler with the same objective and a larger telescope on board.

These space telescopes operate by watching a given field of view that is especially dense with stars, monitoring the brightnesses of many stars simultaneously. Scientists can't know in advance which star has a planet that will transit the star and when that will happen, or which star has no such planet. So they look at many stars, and some of them will yield pay dirt. Another new method of exoplanet hunting is based on an observational consequence of Albert Einstein's general theory of relativity. In this method, scientists also watch a great many stars at once, not knowing where they will find a planet and where they will not. But rather than looking for a very small, brief drop in a star's brightness that is caused by its own planet passing in front of it, they are looking for a fairly large, temporary increase in the apparent brightness of one of the stars, which is caused by the presence of a planet of *another* star, much closer to Earth, passing across the line of sight (or nearly so) to the more distant star. The gravity of the exoplanet concentrates the light of the background star by a process called gravitational microlensing, a bit like a magnifying glass that concentrates the light of the Sun so well that it can be used to start a fire. The planet's own star also causes microlensing of the background star, and the astronomer has to analyze the combined effects of the two brightenings to identify and characterize the planet.

A nice twist in the history of exoplanet research came on August 7, 2000, when William Cochran of the McDonald Observatory at the University of Texas told a symposium at an IAU General Assembly in Manchester, UK (where Andrew Lyne is now an honorary professor of physics and still studies pulsars), that his team had discovered a Jupiter-class planet around Epsilon Eridani, where Bruce Campbell, Gordon Walker, and Stephenson Yang had strongly suspected a planet in 1987. As noted earlier in this chapter, the Campbell findings had gone unconfirmed and unaccepted at the time. But the successful discovery of the Epsilon Eridani planet in 2000 was based not only on new data, but on some of the old Campbell data, which turned out to be useful for planet hunting after all.

Controversy continues in exoplanetary studies, as when NASA erroneously claimed the first direct image of an exoplanet (a photo from the Hubble Space Telescope was misinterpreted) and when some subsequent images from other telescopes also were disputed.

But with the passage of time, heated arguments are now far outweighed by scientific results that reflect a growing consensus that we are now able to detect and study exoplanets using advanced astronomical techniques. These results include recent photographs that *do* show images of individual exoplanets and even two exoplanets that are orbiting the same star. At one time, Marcy and Butler were the only astronomers reporting exoplanet detections to packed halls at American Astronomical Society meetings. Now hundreds of astronomers have joined this field, and there are sessions at these meetings devoted entirely to new findings on exoplanets.

At least one exoplanetary system with five known planets has been discovered, and astronomers around the world are competing to find smaller and smaller planets of sunlike stars. Their main aim is to find planets as small as Earth and then to find such an Earth that is located in the so-called habitable zone around its star. The habitable zone is the region around a star where, if a planet exists that holds water, it isn't so hot that the water evaporates or so cold that the water is frozen all year round. Many think such a planet could hold life. As Senator Jake Garn said after a ride in orbit aboard the space shuttle, "I gazed out into space, and it was clear to me that there had to be life out there." Someday astronomers will find it.

CHAPTER 11

What's Next
for Pluto?

i contain a number of things which i am trying to forget
[said] the universe

—Don Marquis, *the lives and times of archy and mehitabel*, 1940

Astronomy, like the universe itself, is in a state of ever-quickening expansion, and anything that is set down in printed form encounters the near certainty of being out-of-date by the time it reaches library shelves. We've seen in previous chapters how concepts of what is and what isn't a planet altered over the years, often as the result of an accidental discovery, a new technology, or a new way of looking at existing data.

Moreover, it is arguably harder to write history while it is happening than it is to foretell the future. As we write this final chapter, the hubbub over Pluto and the 2006 IAU definition of "planet" is still very much with us. Astrophysicist Neil de Grasse Tyson, an eloquent spokesperson for astronomy, has been making the rounds of TV talk shows, promoting his popular book, *The Pluto Files*, which describes the reasons for and public reaction to the recent "demotion." Public fascination with the subject, judging by enthusiastic reactions to Tyson's book, has hardly diminished since the IAU made Pluto a dwarf planet. And on the technical side of the issue, astronomers still are embroiled in controversy. In August 2008, professional astronomers in the United States convened a meeting of researchers and educators to debate the IAU decision, a prelude to a debate that may be continued internationally at the General Assembly of the IAU in Rio in August 2009.

Though both of the authors of this book are interested observers of the ongoing controversy, neither of us considers himself a specialist on the solar system, and we can't claim to be the ultimate judges of what has happened to Pluto and what the current controversy will look like a decade from now. But we can set down our informed opinions and make some educated projections about where current research on Pluto and other objects at the outermost margin of the solar system will lead us.

LARRY MARSCHALL: MY POSITION ON PLUTO

At the IAU General Assembly in Prague on August 24, 2006, I cast a vote in favor of the current IAU definition of a planet. So, if you are an ardent Pluto partisan, you may accuse me of "demoting" Pluto from the ranks of the planets. But my reading of the situation was certainly not colored by malice. I was merely affirming what astronomers had known for over a decade: Pluto had already, de facto, lost its unique status in the outer solar system as a result of discoveries of larger objects in orbit beyond Neptune.

For many people—astronomers and non-astronomers alike—the IAU vote has become an emotional issue. So let me repeat myself: it wasn't that I didn't like Pluto. In fact, despite being so small, cold, and distant, Pluto ranks among my top 10 favorite solar system objects! But in matters of science, we can't play favorites, and my judgment was that Pluto was significantly different in nature and origin from the other eight planets and so similar to like-sized objects in the outer solar system that it deserved to be considered a member of a new class of objects.

Perhaps because I have little professional stake in planetary astronomy, the precise wording of the IAU definition of "planet" was not really critical to me. Of course the definition had to make sense and be applicable to the objects in question. But as I cast my vote in Prague, I could easily have accepted a number of other broad formulations, as long as they recognized Pluto's kinship to other trans-Neptunian objects. Since August 2006, I've had time to reflect, and I've participated in a number of discussions with fellow astronomers. My view hasn't changed much: the Pluto debate as I see it is largely over semantics, not science.

The purpose of the IAU resolution, after all, was to make it easier for the IAU to conduct the business of assigning names to objects in the solar system—nothing more. Had there not been a problem of how to name new trans-Neptunian bodies like 2003 UB$_{313}$ (a.k.a. Xena and now officially Eris), the IAU might never have gotten into the tricky business of defining the word "planet." Were large trans-Neptunian bodies to be named in the tradition of the traditional major planets with classical names? Or were they to be named, as asteroids are, with names chosen at the will or whimsy of their discoverer?

Beyond this, I thought that any "official" definition would hardly affect the actual conduct of science. I place no great stock in the power of the IAU definition to channel the course of scientific research any more effectively than the French Academy has been able to maintain the purity of the French language. Admittedly, the decision stirred controversy among both astronomers and the general public, but I have difficulty in appreciating all the adrenaline and testosterone that has accompanied the debate. How scientists refer to the objects they are studying is determined not by the actions of professional organizations but by common usage. Technically speaking, for instance, the tomato is a fruit, but at the dinner table, most of us treat it as if it were a vegetable. Let the IAU define planets as it will. I am sure that missions to Pluto and Eris will not go unfunded simply because NASA supports "planetary" astronomy but has no division for "dwarf planetary" astronomy. I also do not think that research on extrasolar planets will wither now that objects orbiting stars other than the Sun are in a sort of lexicographic limbo, unable to be fit into the IAU's definition of a planet.

No, astronomers will continue to call planets as they see them, and if they find themselves—as they do today—with a large number of objects that seem distinct from what they called planets in the past, they will find their own common terms to describe them. Marc Buie summed it up neatly on his Web site recently: "The facts or truths we discover in science come from thinking about the problem and developing a consensus through the scientific method that guides our research. Perhaps the lack of consensus really tells us that we don't know enough yet to define what a planet is." (But Buie told Maran recently that he still strongly believes that Pluto is a planet.)

Of course, the IAU could have finessed controversy altogether by simply staying out of the business of creating definitions. In hindsight, perhaps, this might have been the wisest course of action. As a rule, in fact, the IAU does not assume the responsibility of defining all—or even most—of the technical terms used by astronomers. The officers of the organization could have simply assigned the naming of trans-Neptunian objects to one committee or another, without providing any rationale other than convenience.

But as it turned out, the Pluto brouhaha has been a bonanza for astronomy. For at least several months after the 2006 decision, Pluto was hot news. Students around the world discussed the planets of the solar system in elementary and high schools, and teachers taught about the new discoveries at the edges of the solar system. A group of fourth graders, according to a planetarium director I spoke to recently, picketed her sky show with placards demanding Pluto's reinstatement. And people who never gave a thought to astronomy in the past were posting impassioned analyses to blogs and chat rooms. Perhaps, as the old saying goes, you never miss something till it's gone!

Of course, Pluto is not gone. After a decade or so, I suspect, classes will happily be learning of Pluto and Eris as members of the

Kuiper Belt, and the controversies over the IAU definition will be forgotten. Astronomers will continue to discuss the fine points of planetary classification but largely in calm and measured tones. In the 22nd century, no doubt, the Pluto flap will be a curious historical footnote, like the controversies over whether to name Uranus after King George or the debate over the existence of Vulcan. By then, I am certain, Pluto will only be remembered as the first in a remarkable population of frigid worlds at the outer edges of our solar system.

STEVE MARAN: MY POSITION ON PLUTO

Pluto is a planet! That's my position. Astronomers called it a planet first in 1930, but the concept was soon embedded in popular consciousness in literature and the arts and, above all, in dictionaries. The first edition of *The American Heritage Dictionary of the English Language* defined "Pluto" as a planet (and also as the God of the Underworld). Dictionaries are based on usage. We were taught in school not to use "ain't," but that dictionary defines "ain't" and also decries the use of the word, especially in written English, devoting a paragraph to the issue. Nevertheless, almost everyone uses "ain't" regardless of what grammarians decide. It's part of our culture. So is "Pluto," the ninth planet.

The noted expert on asteroids and comets William F. Bottke hit the nail right on the head in 2007 when he wrote of the IAU decision making in Prague, where he had participated, that "a messy problem was cleaned up by the creation of a new category of objects called 'dwarf planets.'" The "messy problem," as we explained in Chapter 2, was that astronomers were beginning to find other objects beyond Neptune that are comparable to Pluto or even larger, and the list of planets in the solar system was threatening to grow much longer.

That seemed messy, indeed, just as the proliferation of asteroids in the 1800s seemed messy. A lot of astronomers don't want to deal with telling their students about a whole lot of planets in the solar system, and those who write textbooks would like to keep their tables of planetary orbits and physical data (which usually are presented as appendices) to a manageable size. Almost every astronomy textbook used to include tables that listed all known moons of Jupiter and Saturn. There were 11 known Jupiter moons according to the first astronomy book that I ever read, *New Handbook of the Heavens* (1950), and I remember the thrill that I experienced a year later when the 12th moon, Ananke, was found. More than 20 years passed before another Jovian satellite was found, but now, thanks to telescopes in space, cameras on interplanetary space probes, and advanced technology, there are 62 known moons of Jupiter (and 61 of Saturn). They don't get tabulated in texts anymore; you look them up on the Solar System Dynamics Web site of the Jet Propulsion Laboratory. The majority of these moons are tiny compared to the larger and longer-known moons. The Jovian moons are mostly under a few miles in diameter; one is even smaller at less than a mile across. Many of them are in retrograde orbits; they may have been captured by Jupiter instead of formed in a disk of gas and dust around the planet as it was born. Some may have more complicated histories: they may be fragments of larger bodies that were captured by Jupiter and later broken apart by collisions. So the little satellites are very different from the large, long-accepted moons in size, origin, and evolution. Yet, no one says that they aren't moons. There's no "messy problem," because astronomers don't feel that they have to list and name the five dozen plus moons of Jupiter (or Saturn) in dealing with students and the public.

Is "messy problem" really a valid criterion in determining how to define a scientific term? Can't we name and list the first 10 or so largest or most historically known planets and refer the reader to a Web page where all the planets of the solar system are tabulated, just as exoplanets are carefully listed on certain existing Web pages? Professors teach undergraduates about exoplanets nowadays and never think that all 300 and more of them need to be listed in the back of the textbook.

It is scientifically meaningful to divide planets into categories according to their physical nature. The rocky Mercury, Venus, Earth, and Mars are called terrestrial planets. Bloated Jupiter, Saturn, Uranus, and Neptune are "gas giants." It's physically descriptive to call

Pluto and like objects "dwarf planets" or "ice dwarfs," as some scientists prefer. But it makes little sense to declare that a dwarf planet is not a planet. It seems evident that some of those astronomers involved in the debate and voting at the IAU General Assembly at Prague were emphatic in their insistence that Pluto not only be reclassified as a dwarf planet, but also that dwarf planets be excluded from the category of planets because they resented the public fuss over Pluto. The lay world of amateur astronomers and the general public was treading on the hallowed ground so jealously guarded in order that only PhDs and professors of astronomy may enter.

Pluto is still the people's planet and, if anything, more popular as an underdog than ever before. Recently my wife and I enjoyed a visit from Professor Kevin McCartney, the creator of the Maine Solar System Model and the originator of Planet Head Days (as told about in Chapter 1). His hair was beginning to thicken; it had been a few weeks since his head was shaved and painted at the third annual Planet Head Day. (See a video clip on Planet Head Day on YouTube at http://www.youtube.com/watch?v=shZP7A6Y9yY.) McCartney brought us a unique gift. We are now the only ones on our block with a "spare Pluto," a little ball painted to reproduce the Hubble Space Telescope map of Pluto. McCartney commissioned a slew of spare Plutos, because there's one continual problem at the 95-mile-long Maine Solar System Model: people love Pluto, and they keep taking it home with them. McCartney had Pluto encased in glass to discourage theft, but Plutophiles just break the glass and take it away. At last word, no other planet had been stolen in Maine.

Someday there will be a reasoned, logical, and scientific discussion about how to define "planet." In the end, I think fair-minded experts will establish that objects like Pluto are a specific type of planet, and they will include exoplanets, the most numerous planets that we know, in a common classification system with the planets of all sizes and types that exist in our solar system. Until that fine day, Pluto lovers, keep hope alive. I do, and meanwhile, I've got our legally acquired spare Pluto stashed away in a very safe place. I have not revealed its whereabouts to my coauthor.

EXPANDING HORIZONS

Meanwhile, the exploration of the outer solar system continues, unperturbed by occasional rumblings over definitions and classifications rattling the third rock from the Sun. Astronomers like Marc Buie continue to observe Pluto, hoping to learn as much as they can about its surface and its atmosphere using the faint light collected by telescopes on Earth. Michael Brown and a host of others (including Buie) continue to search for additional trans-Neptunian bodies, among which may be dozens or even hundreds of worlds as big or bigger than Pluto.

As they patiently collect data here on Earth, astronomers eagerly look forward to July 14, 2015, the day when the New Horizons space mission spearheaded by Alan Stern is scheduled to encounter Pluto. It may not be Bastille Day way out there, but the New Horizons spacecraft will bring about a revolution in our understanding of the outer solar system.

Starting shortly before the July 14 encounter, instruments on the fast-moving probe will observe the surfaces of Pluto and Charon, not just to map them in much finer detail than ever before, but also to pin down the temperature and chemical composition from one area to another on each of the two bodies. They will also investigate topography, exploring highlands, lowlands, ice caps, and any other features. (Are there craters? If so, are they from impacts or cryovolcanism?) They'll look for more small moons, besides Nix and Hydra, or even a ring or rings too dim to see from Earth. They'll gather more information about the atmosphere of Pluto, and they'll search for any wisp of atmosphere that might exist on Charon. The probe bears instruments that observe and measure visible, infrared, and ultraviolet light and radio waves and also is equipped to measure the solar wind, plasmas, and energetic particles around Pluto.

Pluto has never been visited by a spacecraft before. Judging by our experience when planets like Jupiter and Saturn were visited for the first time, our understanding of Pluto and its moons should undergo a drastic change. The flood of new discoveries will shed light on the origin of this strange body and its relation to its neighbors in the solar system. And there will no doubt be lots of surprising features, some entirely unanticipated, which will keep astronomers busy answering new questions and engaging in new scientific debates.

Then, if all goes well, the New Horizons spacecraft will move on toward an even more distant horizon in the Kuiper Belt. Though specific targets have not yet been chosen, the mission timeline calls for one or two close encounters with trans-Neptunian objects beyond Pluto starting in 2016. Astronomers are hoping to gather enough information about what's out there to make a better-informed choice by 2015, so that the spacecraft can be retargeted to visit the most interesting and accessible worlds in that great beyond. What we will learn from this first encounter with other Kuiper Belt objects we cannot tell, but whether New Horizons reinforces the notion that Pluto is one of the trans-Neptunian family of objects or they appear completely different in crucial respects, the information is bound to be fascinating, and the images of these most distant worlds will likely become iconic.

When New Horizons blasted off on January 19, 2006, the controversy over Pluto's status as a planet had not yet come to a head at the August General Assembly of the IAU. But, as we have described in this book, there had been heated controversies over planets in the centuries that preceded it. We have mentioned many cases in which tempers flared and reputations were tarnished. But it is our experience—both from reading the historical record and from our personal knowledge of those involved in modern controversy—that the disputes, in large part, were carried on in good faith and that errors of fact or interpretation, where they occurred, were honest. Controversy, for better or worse, is the way good science flourishes. If we knew all the answers, or if it were easy to learn them, there would be fewer disputes, and the practice of science would be much less exciting.

So as the New Horizons spacecraft speeds onward to a rendezvous with Pluto—however that distant world is "officially" classified—we salute the mission and the centuries of scientific progress and controversy that made it possible. We two authors may disagree respectfully on whether Pluto is a planet, but we are united in the belief that New Horizons will be a great success.

ACKNOWLEDGMENTS

We thank Lars Lindberg Christensen, Press Officer of the International Astronomical Union, for his interest in this project, his hospitality to LAM during the 2006 General Assembly in Prague, and for sending us a valuable compendium of media planning documents, IAU resolutions, press releases, and other materials assembled after that meeting.

The authors are especially grateful to their agent, Skip Barker, and to BenBella's Publisher, Glenn Yeffeth, for their strong encouragement, which brought this project to a successful conclusion. We also thank the dedicated staff at BenBella Books for their work in editing, design, and production.

Laurence A. Marschall: For sharing one of his historical studies on the discovery of Pluto, I'd like to thank Dr. David DeVorkin. Over the course of the writing of this book, I've had many informal conversations on Pluto with fellow astronomers who have provided opinions, insights, and background on both the scientific and the political aspects of the subject, among them John Huchra, Owen Gingerich, Don Goldsmith, and Dan Green. Thanks to the librarians at the U.S. Naval Observatory in Washington, D.C., whose remarkable collection provided materials not otherwise available. Thanks to the members of the Physics Department at Gettysburg College, who learned not to knock when I was closeted with my manuscript, and to Joe Medhurst and Heather Quinn of Lancaster, UK, whose New Wood House provided the ideal retreat for a writer during the Fall of 2008. To Gettysburg College, NASA, and the National Science Foundation, who have supported my professional travels, among them my attendance at the IAU General Assembly in Prague, August 2006, and many observing trips to Lowell Observatory. Most of all I'd like to thank my family, Ellen, Emma, and Geoff, for the support and patience that not only made the writing of this book possible, but made it all worthwhile.

Stephen P. Maran: For helpful discussions during the preparation of this book I thank Dr. Marc W. Buie (Pluto research), Prof. David Jewitt (Kuiper Belt), Mr. David H. Levy (public attitudes on Pluto), Prof. Kevin McCartney (Maine Solar System Model and Planet Head Days), and Prof. Vigdor Teplitz (Kuiper Belt dust). I also gratefully acknowledge helpful discussions, conversations, and correspondence on the topics of Pluto, Charon, Barnard's star, and VB 8B in previous years with the late Dr. Robert S. Harrington and with Drs. Alan Stern (Pluto, Kuiper Belt) and Brian G. Marsden (Kuiper Belt) that, preserved in notes and files, directly assisted the preparation of the manuscript of *Pluto Confidential*.

I especially thank my wife Sally Maran for typing four and a half chapters of the manuscript, making editorial improvements while doing so, after a medical problem forced me to compose chapters in longhand rather than on a keyboard. Dr. Richard T. Fienberg, at considerable sacrifice, took over most of my duties as Press Officer of the American Astronomical Society during the same period, for which I will always be grateful. I thank Sally, our daughters Enid and Elissa, and my grandson Michael Scott Bean for their love and support.

NOTES ON SOURCES

CHAPTER 1: SUMMER IN PRAGUE

Since much of the material we describe here is ephemeral or late-breaking, there are few general references yet published, but we highly recommend two highly readable books on the Pluto controversy, *The Pluto Files,* by astronomer Neil deGrasse Tyson (New York: W. W. Norton, 2009), and *The Hunt for Planet X: New Worlds and the Fate of Pluto* by science journalist Govert Schilling (New York: Copernicus Books, 2009). In addition, there is a wealth of material in blogs and Web archives, just a Google away.

—California 2006 state Assembly bill, HR 36: http://www.leginfo .ca.gov/pub/05-06/bill/asm/ab_0001-0050/hr_36_bill_ 20060824_introduced.pdf

—City of Madison Legislative File Number 04419: http://legistar .cityofmadison.com/DetailReport/Reports/Temp/5420091 52239.pdf

—New Mexico 2007 House Joint Memorial 54: http://legis.state. nm.us/Sessions/07 Regular/memorials/house/HJM054.htm

—About Rancho Hidalgo: David H. Levy, http://www.skyand telescope.com/community/skyblog/observingblog/38821737. html

— "We, as planetary scientists and astronomers, do not agree with the IAU's definition of a planet, nor will we use it. A better definition is needed.": http://www.ipetitions.com/petition/planetprotest/

— "The New Horizons project, like a growing number of the public, and many hundreds if not thousands of professional research astronomers and planetary scientists, will not recognize the IAU's planet definition resolution of Aug. 24, 2006.": Alan Stern, http://pluto.jhuapl.edu/overview/piPerspective.php?page=piPerspective_09_06_2006

— "the controversy over a sole correct definition of 'planet,' and whether Pluto falls within it, is unwarranted from a scientific perspective.": Jeffrey Parsons and Yair Wind, "A question of class," *Nature*, 455, 1040–1041, October 23, 2008.

— "who has the social mandate to alter the definition of something as fundamental as a planet." Mark Bullock, http://www.spaceref.com/news/viewpr.html?pid=20724

— "the process that astronomers followed was embarrassing, unkempt, and overly politicized.": David A. Weintraub, "Pluto, We Hardly Knew Ye," *The Chronicle of Higher Education*, 53, No. 31, p. 20 (April 6, 2007).

— "To pluto is to demote or devalue someone or something, as happened to the former planet Pluto when the General Assembly of the International Astronomical Union decided Pluto no longer met its definition of a planet.": http://www.americandialect.org/index.php/amerdial/plutoed_voted_2006_word_of_the_year/ and http://www.americandialect.org/Word-of-the-Year_2006.pdf

— "Adieu, Poor Pluto, Sent to the Doghouse," *The Age*, Melbourne, August 28, 2006.

— "With Pluto gone, which of us will be next?" *The Star*, Toronto, September 2, 2006.

— "Going 'round and 'round on defining Pluto," Chet Raymo, *Boston Globe*, August 28, 2006.

— "Astronomers goofed on Pluto," *The Times Union*, Albany, August 30, 2006.

- —"Pluto's characterization has bugged me since I was an early teen," Proesterchen, http://blogs.discovermagazine.com/badastronomy/2006/08/24/breaking-news-pluto-not-a-planet/#comment-19813
- —"Pluto rocks, in a vacuous void/ Grandfather Pluto, he's no asteroid.": http://www.myspace.com/subplotaplutorocks
- —"They met in Prague and voted/Now Pluto's been demoted/Oh, Pluto's not a planet anymore.": http://jeffspoemsforkids.com/s1.php?id=22
- —The Maine Solar System Model: http://www.umpi.maine.edu/info/nmms/solar/
- —Planet Head Days: http://www.youtube.com/watch?v=shZP7A6Y9yY
- —"Home of the largest complete model of the solar system in the world," http://www.lakeview-museum.org/pdfs/solar_system.pdf
- —"When Pluto is in the IVth house and inharmoniously aspected, he creates fanaticism, intolerance, phantasy, isolation.": Fritz Brunhuebner, Pluto (1934), English translation by Julie Baum, revised April 1971, p. 57, Washington, DC: The American Federation of Astrologers.
- —The Pluto demotion by the IAU "is substantially irrelevant to astrology": Rob Tillett, http://www.astrologycom.com/pluto.html
- —Although Pluto has been reclassified as a dwarf planet, "astrologers still reckon it to be a powerful force on a collective and individual level.": Molly Hall, http://astrology.about.com/od/advancedastrology/p/Pluto.htm
- —"Astronomers behaving badly.": David Finley, private communication to SPM, August 24, 2006.

CHAPTER 2: THE GREAT PLUTO DEBATE AND THE GREAT PLUTO DEBATERS

Much of this chapter relies on personal notes and informal communications, but the books by Neil deGrasse Tyson and Govert Schilling, noted in the sources for Chapter 1, are excellent background reading.

—An American textbook taught that there were 11 planets: Denison Olmsted 1848, *A Compendium of Astronomy*, p. 169, New York: Collins and Brother.

—The number of planets had grown to "about ninety" by one reckoning: "A Connecticut Pastor," 1867, *Parish Astronomy*, Boston: Nichols & Noyes, p. 93.

—"Ninth planet of the solar system.": Otto Struve and Velta Zebergs, 1962, *Astronomy of the 20th Century*, endpaper, New York: Macmillan.

—Definition of "planet" as a "Large body that orbits the Sun or other star [and] that shines only by light reflected from the Sun or star.": Jeff Hester, David Burstein, George Blumenthal, Ronald Greeley, Bradford Smith, Howard Voss, and Gary Wegner, 2002, p. G-12, New York: W. W. Norton & Company.

—Is Pluto a "Tiny Planet or Gigantic Comet?": *ibid*, p. 269.

—"If discovered today, they [Pluto and its largest moon, Charon] would certainly be considered especially large KBOs rather than a planet and its moon.": *ibid*, p. 271.

—"Is it not likely that in Pluto there has come to light the *first* of a *series* of ultra-Neptunian bodies, the remaining members of which still await discovery but which are destined eventually to be detected?": Frederick C. Leonard, 1930, "The New Planet Pluto," *Astronomical Society of the Pacific Leaflet,* Number 30, p. 124.

—Before 1992, the only "observed member [of the Kuiper Belt] was Pluto, misleadingly given planetary status for a host of mostly socio-scientific reasons." And: "Our study of the structure of the solar system could have been advanced by many decades. . . .": David Jewitt, 2008, "Kuiper Belt and Comets: An Observational Perspective," in David Jewitt, Alessandro Morbidelli, and Haike Rauer, *Trans-Neptunian Objects and Comets*, Berlin: Springer, pp. 1–78 (see p. 22–23).

—"Young man you are wasting your time.": John Noble Wilford, 1980, "50 Years After Discovery, Pluto Is Shedding Its Mystery," *New York Times*, p. C-1 (February 19, 1980).

—Tyson is "America's champion spokesman for space.": Dava Sobel, 2009, in Neil deGrasse Tyson, *The Pluto Files*, New York: W. W. Norton & Company, endorsement on dust jacket.

—"I blame it [the public American uproar in planet Pluto's defense] on the dog because people in Europe don't behave this way," Neil deGrasse Tyson, 2009, televised remarks, quoted in *APS News*, 18, No. 3, p. 2.

—Members of the IAU's Planet Definition Committee sat slumped in their chairs as critics lined up at microphones to "denounce the definition in tones ranging from offended to furious.": Jenny Hogan, 2006, Diary of a planet's demise," *Nature*, 442, pp. 966–967.

—At one point in the discussion of the Committee's recommendations, Milani "literally screamed": Govert Schilling, 2006, "Underworld Character Kicked Out of Planetary Family," *Science*, 313, pp. 1214–1215.

—"Astronomers behaving badly.": David Finley, 2009, private communication to Maran (August 24, 2009).

—"Brian and I explicitly discussed the question of dual classification—he was in favor of it, and brought it up with me.": Michael F. A'Hearn, telephone interview with Maran (March 19, 2009).

—"I gave a talk that was mostly about *Archaeopteryx* and the more than 100 years of argument about the transition between a dinosaur and a bird.": Michael F. A'Hearn, *ibid*.

—The Planet Definition Committee "had vigorous discussions of both the scientific and the cultural/historical issues." Some members lost sleep "worrying that we would not be able to reach a consensus.": Owen Gingerich, 2006, in News Release IAU0601: The IAU draft definition of "planet" and "plutons," http://www.iau.org/public_press/news/release/iau0601/

—"A planet is a celestial body that (a) has sufficient mass for its self-gravity to overcome rigid body forces so that it assumes a hydrostatic equilibrium (nearly round) shape, and (b) is in orbit around a star, and is neither a star nor a satellite of another planet.": International Astronomical Union, 2006, News Release IAU0601: The IAU draft definition of "planet" and "plutons," http://www.iau.org/public_press/news/release/iau0601/

—The Planet Definition Committee "so far as I know never solicited input," and, "The IAU membership heard about [the proposed definition] about the same time that the press did.": Michael F. A'Hearn, *loc. cit.*

—(1) A planet is a celestial body that (a) is in orbit around the Sun, (b) has sufficient mass for its self-gravity to overcome rigid body forces so that it assumes a hydrostatic equilibrium (nearly round) shape, and (c) has cleared the neighbourhood around its orbit. AND (2) A dwarf planet is a celestial body that (a) is in orbit around the Sun, (b) has sufficient mass for its self-gravity to overcome rigid body forces so that it assumes a hydrostatic equilibrium (nearly round) shape, (c) has not cleared the neighbourhood around its orbit, and (d) is not a satellite. AND (3) All other objects orbiting the Sun shall be referred to collectively as "Small Solar System Bodies.": International Astronomical Union 2006, News Release IAU0603: IAU 2006 General Assembly: Result of the IAU Resolution votes, http://www.iau.org/public_press/news/release/iau0603/

—There was no official record made of the vote which the IAU later described as "a great majority.": International Astronomical Union, *ibid.*

—A footnote stated: "The eight planets are: Mercury, Venus, Earth, Mars, Jupiter, Saturn, Uranus, and Neptune.": International Astronomical Union, *ibid.*

—Another decision was made by a vote of 237 to 157, with 17 astronomers abstaining, that, "The IAU further resolves: Pluto is a 'dwarf planet' by the above definition and is recognized as the prototype of a new category of trans-Neptunian objects." International Astronomical Union, *ibid.*

—The Illinois State Senate adopted a resolution which stated that "Pluto was unfairly downgraded to a 'dwarf' planet.'": http://www.ilga.gov/legislation/96/SR/PDF/09600SR0046lv.pdf

—"No proposal to change the status of Pluto as the ninth planet in the solar system has been made by any division, Commission or Working Group of the IAU responsible for solar system science. Accordingly, no such initiative has been considered by the Officers or Executive Committee, who set the policy of the IAU itself.": Johannes Andersen 1999, The Status of Pluto: A Clarification, IAU Press Release 01/99 (February 3, 1999).

—"On one exhibit wall, I noticed the glaring absence of Pluto among the other planets of our own and other planetary systems.": Stephen P. Maran, 2000, "Selling Astrophysics in a Crystal Palace," *Sky and Telescope*, 99, No. 5, pp. 46–47.

—The *New York Times* ran a front page story: Kenneth Chang, "Pluto's Not a Planet? Only in New York," *New York Times* (January 22, 2001).

—Mark Sykes argued with Tyson over Pluto: Kenneth Chang, "Icy Pluto's Fall From the Planetary Ranks: A Conversation," *New York Times* (February 13, 2001).

CHAPTER 3: CONTENTIOUS PLANETS: THE EARLY HISTORY OF PLANETARY DISPUTES

For further reading on Galileo, refer to our own book *Galileo's New Universe*, by Stephen P. Maran and Laurence A. Marschall (Dallas: BenBella, 2009), which contains a reading list of additional primary and secondary references. A recent and very readable short biography of the Herschels is Michael Lemonick's *The Georgian Star* (New York: W. W. Norton and Company, 2008), and for a more scholarly treatment, *The Herschel Partnership*, by Michael Hoskin (Cambridge: Science History Publications, 2003).

—"We have," he wrote in *Sidereus Nuncius*, "an excellent and splendid argument for taking away the scruples of those who . . ." Albert Van Helden, *Sidereus Nuncius, or the Sidereal Messenger,* Chicago: University of Chicago Press, 1989, p. 84.

—"It is a marvelous thing, a stupendous thing; whether it is true or false, I know not." Letter to Kepler by Martin Horky in Stillman Drake, *Galileo Studies*, Ann Arbor: University of Michigan Press, 1970, p. 132

—"I do not wish to approve of claims about which I do not have any knowledge . . ." quoted in Quodlibeta: http://bedejournal .blogspot.com/2006/11/who-refused-to-look-through-galileos .html

—"What shall we make of all this, shall we laugh or shall we cry?" Wade Rowland, *Galileo's Mistake*, New York: Arcade Publishing, 2003, p. 95

—"I found that these fathers, having finally recognized the truth of the new Medicean planets, have been observing them . . ." in James Lattis, *Between Copernicus and Galileo: Christopher Clavius and he Collapse of Ptolemaic Astronomy*, Chicago: The University of Chicago Press, 1994, p. 187

—"I was so delighted with the subject that I wished to see the heavens and Planets with my own eyes thro' one of those instruments." In Constance Lubbock, editor, *The Herschel Chronicle*, Cambridge: Cambridge University Press, 1933, p. 59.

—"It was agreed that he should also show me the manner in which he had proceeded with grinding and polishing his mirrors." In Joseph Ashbrook, *The Astronomical Scrapbook*, Cambridge: Cambridge University Press, 1984, p. 128.

—"Heaped up with globes, maps, telescopes, reflectors, &c., under which his piano was hid." In Lemonick, *The Georgian Star*, p. 69.

—"a curious either nebulous star or perhaps a comet." *Ibid*, p. 72.

—"I was enabled last night to discern a motion . . . which, as well as from its agreeing with the position . . ." In Lubbock, *op. cit.*, p. 79.

—"I am constantly astonished at this comet, as it does not resemble any one of those I have observed, whose number is eighteen." In Lubbock, *op .cit.*, p. 87.

—"I can now say that I absolutely have the best telescope that were ever made." In Mark Littman, *Planets Beyond*, New York: John Wiley & Sons, Inc., 1988, p. 9.

—"This is an evident mistake," Herschel countered; "In the regular manner I examined every star of the heavens. . . ." In Lubbock, *op.cit.*, p. 79.

—"Among opticians and astronomers nothing now is talked of but what they call my Great discoveries. . . ." in Lemonick, *op. cit.*, p.87.

—"Our nimble neighbors, the French, will certainly save us the trouble of Baptizing it." In Lubbock, *op. cit.*, p. 95.

—"In the fabulous ages of ancient times the appellations of Mercury, Venus, Mars, Jupiter, and Saturn, were given to the planets. . . ." William Herschel, *Philosophical Transactions*, 73, 1783, p. 324.

—"You know perhaps, that I was the first person in Germany to see the new star. . . ." In Lubbock, *op.cit.*, p. 123.

CHAPTER 4: CERES AND THEORIES: THE SEARCH FOR PLANETS BEGINS

An eminently readable and exhaustively researched popular account of asteroid discovery (and planetary discovery in general) is Mark Littman's *Planets Beyond: Discovering the Outer Solar System* (New York: John Wiley & Sons, 1988). Though the chapters on satellite exploration have been partly superseded by two decades of exploration of Jupiter and Saturn, the historical chapters are still lucid and authoritative.

—"Can one believe that the Founder of the Universe left this space empty? Certainly not." In Littman, *op.cit.*, p. 16.

—"I have announced this star as a comet. . . ." In Littman, *op.cit.*, p. 18.

—". . . the news from Palermo may be said to have converted him . . ." Agnes Clerke, *A Popular History of Astronomy During the Nineteenth Century (Fourth Edition)*, London: Adam and Charles Black, 1908, p. 74.

—". . . confirmation of that beautiful progression . . ." Stanley Jaki, "The Early History of the Titius-Bode Law," *American Journal of Physics*, 40, 1972, p. 1019.

—"While the displacements of the planets suggest an arithmetic progression . . ." G. W .F. Hegel, *De Orbitis Planetarum*, Jena, 1801, translated from the Latin original by David Healan, http://www.hegel.net/en/v2133healan.htm

—"I have the full right to name it. . . ." In Clifford Cunningham, *The First Asteroid*, Surfside, Florida: Star Lab Press, 2001, p. 49.

—"Q: How many primary planets are there? . . ." Asa Smith, *An Abridgement of Smith's Illustrated Astronomy designed for the use of Junior Classes in the Public or Common Schools in the United States*, New York: Cady & Burgess, 1849, p. 10.

CHAPTER 5: NEPTUNE'S DISPUTED DISCOVERY

W. M. Smart's essay "John Couch Adams and the Discovery of Neptune," published as *Occasional Notes of the Royal Astronomical Society* in 1947 to commemorate the discovery a century earlier, is a very complete and readable source, as is Morton Grosser's *The Discovery of Neptune* (Cambridge: Harvard University Press, 1962). Mark Littman's *Planets Beyond* is also very good on the subject, and perhaps more easily obtainable at the local library.

—". . . movement of the age . . ." in Agnes Clerke, *op. cit.*, p. 78.

—". . . an entire revolution of the moon's nodes. . . ." Entry on "Lemonnier, Pierre Charles," in *Encyclopedia Britannica*, 11th edition.

—". . . If it were certain that there were any extraneous action. . . ." W. M.Smart, *op.cit.*, p. 8.

—"Formed a design, in the beginning of this week," John Couch Adams, "The History of the Discovery of Neptune" in Harlow Shapley and Helen E. Howarth, *A Source Book in Astronomy*, New York: McGraw-Hill Company, Inc., 1929, p. 246.

—"It was so novel a thing to undertake observation in reliance on merely theoretical deductions . . ." Smart, *op. cit.*, p. 28.

—". . . bolder and more energetic man. . . ." *ibid.*, p. 28.

—". . . now I felt no doubt of the accuracy of both calculations. . . ." *ibid.*, p. 26.

—". . . I am asking, almost at a venture. . . ." *ibid.*, p. 27.

—"We see it as Columbus saw America. . . ." *ibid.*, p. 29.

—"The planet whose position you have pointed out *actually exists*. . . .", J. P. Nichol, *The Planet Neptune: An Exposition and History*, Edingburgh: John Johnstone, 1848, p. 89.

—". . . discovered a star with the tip of his pen. . . ." Richard Baum and William Sheehan, *In Search of Planet Vulcan*, New York: Plenum Trade, 1997, p. 2.

—"The entire annals of Observations probably do not elsewhere exhibit. . . ." J. P. Nichol, *op. cit.*, p. 90.

—"The remarkable calculations of M. Le Verrier. . . ." W. M. Smart, *op. cit.*, p. 36.

—". . . the merit of the clandestine researches of M. Adams. . . ." *ibid.*, p. 33.

—"After four days of observing, the planet was in my grasp. . . ." Mark Littman, *op. cit.*, p. 50.

—"Would it not be worth while to look at it with a higher power?" W. M.Smart, *op. cit.*, p. 31.

—" You are to be recognized. . . ." *ibid.*, p. 33

—"Heartily do I wish. . . ." *ibid.*, p. 33.

—"I may mention that I am not yet in Orders," *ibid.*, p. 34.

—"Since Copernicus . . . nothing (in my opinion) so bold, and so justifiably bold. . . ." *ibid.*, p. 38.

—"You were accused, not only of unreasonable incredulity and apathy. . . ." *ibid.*, p. 38.

—"I was abused most savagely both by English and French." George Biddle Airy, *Autobiography*, Cambridge: Cambridge University Press, 1896, p. 181.

—"To do it justice it is candid. . . ." W. M. Smart, *op. cit.*, p. 38.

—"I remember . . . being charmed, like everyone else. . . ." *ibid.*, p. 53.

Chapter 6: Vulcan, The Planet That Wasn't

Richard Baum and William Sheehan have written the most complete and accessible account of this phantom planet, *In Search of Planet Vulcan* (New York: Plenum Trade, 1997). For a much more technical history of Vulcan and proposed 19th century solutions to the problematic orbit of Mercury, see N. T. Roseveare, *Mercury's Perihelion from LeVerrier to Einstein* (Oxford: Oxford University Press, 1982).

—"I may compare myself to Saul, who went out to seek his father's asses. . . ." Richard Baum and William Sheehan, *op. cit.*, p. 141.

—". . . absolute conviction that the details of his observations ought to be admitted. . . ." *ibid.*, p. 155.

—"Garibaldi and the weather ceased to interest the Parisians. . . ." *ibid.*, p. 4.

—". . . applauding this second triumphant conclusion. . . ." Robert Fontenrose, "In Search of Vulcan," *Journal for the History of Astronomy*, Volume IV, 1973, p. 146.

—"Described by a contemporary as an astronomer 'of considerable skill,'" Richard Proctor, *Myths and Marvels of Astronomy*, London: Longmans, Green and Co., 1893, p. 317.

—"I am in a position to deny, and to deny positively and absolutely. . . ." Richard Baum and William Sheehan, *op. cit.*, p. 164.

—"A hypothesis or theory is clear, decisive, and positive, but it is believed by no one. . . ." Harlow Shapley, *Through Rugged Ways to the Stars*, New York: Scribner, 1969.

—"To see Vulcan through the sun. . . ." Richard Proctor, *op.cit.*, p. 325.

—". . . minute object near the sun," in B. F. Sands, *Reports on Observations of the Total Eclipse of the Sun, August 7, 1869*, Washington: Government Printing Office, p. 180.

—"in reference to the sun and a neighboring star . . . by a method which obviates the possibility of error. . . ." Robert Fontenrose, *op. cit.*, p. 150.

—". . . it is of course now a matter of great regret. . . ." Simon Newcomb, in *Reports on the Total Solar Eclipses of July 29, 1878 and January 11, 1880*, Washington, DC: Government Printing Office, p. 105.

—"Professor Watson of Ann Arbor, Michigan, who is now the most noted. . . ." Richard Baum and William Sheehan, *op. cit.*, p. 209.

—"The planet Vulcan, after so long eluding the hunters. . . ." Robert Fontenrose, *op. cit.*, p. 151.

—"because the searches were made very thoroughly. . . ." Robert Fontenrose, *op. cit.*, p. 152.

—"With few exceptions, only persons otherwise wholly unknown. . . ." Richard Baum and William Sheehan, *op. cit.*, p. 220.

—". . . any such bodies would fail hopelessly. . . ." Robert Fontenrose, *op. cit.*, p. 155.

—"The explanation of the shift in Mercury's perihelion. . . ." *ibid.*, p. 156.

CHAPTER 7: THE PLUTO SAGA BEGINS

The best and most complete popular study on the discovery of Pluto up to the time its discovery is *Planets X and Pluto* by William Graves Hoyt (Tucson: The University of Arizona Press, 1980). For a very entertaining and accessible insight into Clyde Tombaugh's search for "planet X," see his articles "The Discovery of Pluto: Some Generally Unknown Aspects of the Story," in *Mercury*, 16, 1986, p. 66 (part I), p. 98 (part II).

—". . . after thirty or forty years of observing the new planet. . . ." William Graves Hoyt, *op. cit.*, p. 71.

—". . . Donati's comet of 1858 being my earliest recollection. . . ." William Graves Hoyt, *Lowell and Mars*, Tucson: University of Arizona Press, 1976, p. 15.

—"That Mars is inhabited . . . we may consider as certain. . . ." Percival Lowell, *Mars and Its Canals*, New York: The Macmillan Company, 1906, p. 376.

—"I venture to think that his merit as one of our first astronomical observers. . . ." Alfred Russel Wallace, *Is Mars Habitable?*,
London: The Macmillan Company, 1907, p. 99.

—"For so complicated is the problem that all elementary means
of dealing with it. . . ." Percival Lowell, *Memoirs of the Lowell
Observatory*, 1, 1915, p. 3.

—"Owing to the inexactitude of our data, then, we cannot
regard . . ." *ibid.*, p. 103.

—". . . look fairly good for such a chap working all alone. . . ."
William Graves Hoyt, *Planets X and Pluto*, p. 179.

—". . . it was the only planetary observatory I knew of. . . ."
Clyde W. Tombaugh and Patrick Moore, *Out of the Darkness:
The Planet Pluto*, Harrisburg, PA: Stackpole Books, 1980, p.
25.

—". . . He has a good attitude. . . ." William Graves Hoyt, *Planets
X and Pluto*, p. 181.

—". . . It was better than pitching hay back on the Kansas
Farm. . . ." Clyde W. Tombaugh, "The Predictions and Dis-
covery of the Ninth Planet, and the Extensive Planet Search,"
Space Science Reviews, 43, 1986, p. 284.

—". . . The experience was an intense thrill. . . ." Clyde W. Tom-
baugh, "The Discovery of Pluto," in Harlow Shapley, *Source
Book in Astronomy, 1900-1950*, Cambridge: Harvard University
Press, 1950, p. 73.

—". . . this could be very hot news . . ." Clyde W. Tombaugh,
"The Discovery of Pluto: Some Generally Unknown Aspects of
the Story, Part II," in *Mercury,* 16, 1986, p. 98.

CHAPTER 8: THE EXPLORATION OF PLUTO

Good general information on Pluto since its discovery can be found
in *Pluto and Charon: Icy Worlds on the Ragged Edge of the Solar System*,
by Alan Stern and Jacqueline Mitton (New York: John Wiley and
Sons, Inc., 1998).

—Pluto's "diameter is 0.46 times the earth, midway between
Mars's and Mercury's" diameters.: Gerard P. Kuiper, 1950,

"The Diameter of Pluto," *Publications of the Astronomical Society of the Pacific*, 62, pp. 133–137.

—"To his [Christy's] surprise, he immediately noted that the images of Pluto were elongated—all the exposures on 13 and 20 April showed a faint southerly extension, and those on 2 May showed a faint northerly extension.": Joseph Marcus and Richard Schmidt, 1978, *Comet News Service*, No. 78-5, pp. 1–2.

—"Frankly we were not interested in Pluto, and the observations were taken primarily because they represented very little additional effort, and could eventually be useful to someone, somewhere, sometime.": J. Derral Mulholland, 1978, Pluto's Neighbor, *Science*, 201, p.867.

—"The last time we had a press conference here was 101 years ago." Anonymous, 1978, "Moon Believed Found for Pluto," *Science News*, 114, p.36.

—"It beat pitching hay": John Noble Wilford, 1980, "50 Years After Discovery, Pluto Is Shedding Its Mystery," *New York Times,* p. C-1 (February 19, 1980).

—When planetary astronomers first learned of Andersson's prediction of the upcoming occultations, "the salivating was reminiscent of Pavlov's dogs.": Alan Stern and Jacqueline Mitton 1998, *Pluto and Charon: Icy Worlds on the Ragged Edge of the Solar System*, New York: John Wiley and Sons, p. 74.

—"The data may be attributed to a direct detection of polar caps on Pluto.:" Richard P. Binzel, 1988, "Hemispherical Color Differences on Pluto and Charon," *Science*, 241, pp. 1070–1072.

—"Many astronomers doubted whether Pluto was massive enough to hold an atmosphere," Anonymous, 1988, "Barren planet bares its air," *New Scientist*, 118, No. 1619, p. 45.

—"Incontrovertible evidence for gaseous methane"on Pluto: Alan Stern, 1988, private communication to Maran (e-mail, August 20, 1988)

—"A hugely extended atmosphere—this is caused by Pluto's very low gravity.": Alan Stern 1988, *ibid.*

—"Pluto is no longer the small, rocky planet that didn't fit the pattern of the solar system, it is now the small, fluffy planet that doesn't fit the pattern of the solar system.": Edward C. Krupp, 1978, *Griffith Observer*, 42, No. 10, pp. 9–14.

—"To human eyes, Charon would appear a bland, neutral gray, and Pluto would be reddish.": Richard P. Binzel, 1990, Pluto, *Scientific American*, 262, No. 6, pp. 50–58.

—Pluto is "probably about 75 percent rock and 25 percent ice by mass.": Steve Mueller and William B. McKinnon, 1989, "The Changing Status of Pluto," *The World & I*, 4, No. 1, pp. 274–281.

—Pluto is "a critically important object, a large rock-ice world, and a planet by convention, born at the boundary between the realm of gas giants and the Kuiper belt of comets beyond.": William B. McKinnon, 1993, "An enigma orbiting a puzzle," *Nature*, 365, pp. 209–210.

—"Pluto's atmosphere is dominated by nitrogen or carbon monoxide rather than methane.": S. Alan Stern, David A. Weintraub, and Michel C. Festou, 1993, "Evidence for a Low Surface Temperature on Pluto from Millimeter-Wave Thermal Emission Measurements," *Science*, 261, pp. 1713–1716.

—In May 1994, NASA released an image described as "the clearest view of the distant planet Pluto, and its moon Charon, as revealed by the Hubble Space Telescope." Pluto and Charon were "shown as clearly separate and sharp disks.": http://hubblesite.org/newscenter/archive/releases/solar-system/pluto/1994/17/

—By 1996, NASA was reporting that the orbiting telescope had pictured "nearly the entire surface of Pluto" and revealed that "For the first time since Pluto's discovery 66 years ago, astronomers have at last directly seen details on the surface of the solar system's farthest known planet." And: Buie said that Pluto was now "a world which we can begin to map and watch for surface changes." And: Stern called the lighter areas "as bright as fresh Colorado snow.": http://hubblesite.org/newscenter/archive/releases/solar-system/pluto/1996/09/text/

- —When a veteran Pluto researcher at the Southwest Research Institute in Boulder, Colorado, told us about these expeditions to us in April 2009, he was about to fly to Africa to watch an occultation from Namibia . . . : Marc W. Buie 2009, telephone interview with Maran (April 10, 2009).
- —The atmosphere had "expanded rather than collapsed.": James L. Elliott, et al. 2003, "The recent expansion of Pluto's atmosphere," *Nature*, 424, pp. 165–168.
- —The Sicardy group called the change "a probable seasonal effect": Bruno Sicardy, et al. 2003, "The recent expansion of Pluto's atmosphere," *Nature*, 424, pp. 168–170.
- —They attributed these brightness "spikes" to refraction effects: Jay M. Pasachoff, et al. 2005, "The Structure of Pluto's Atmosphere from the 2002 August 21 Stellar Occultation," *Astronomical Journal*, 129, pp. 1718–1723.
- —The occultation data ruled out the presence of any "significant atmosphere" on Charon. And: The Gulbis team commented that the findings "seem to be consistent with collisional formation for the Pluto-Charon system.": Amanda A. S. Gulbis, et al. 2006, *Nature*, 439, pp. 48–51.
- —"Charon may be a piece of bare rock with a density similar to that of the terrestrial planets or the Moon.": Ernst J. Öpik, 1978, "Charon—the remarkable Satellite of Pluto," *Irish Astronomical Journal*, 13, pp. 198–201.

CHAPTER 9: UNVEILING THE KUIPER BELT

A quite readable introduction to current research on the Kuiper Belt is *Beyond Pluto: Exploring the Outer Limits of the Solar System*, by John Davies (Cambridge: Cambridge University Press, 2001). For a more technical and more current view of the field by some prominent researchers, we recommend *Trans-Neptunian Objects and Comets* by David Jewitt, Alessandro Morbidelli, and Heike Rauer (Berlin: Springer, 2008).

- —!!Seen THIS??: Alan Stern 1992, private communication (e-mail) to Maran (September 15, 1992).

—David Jewitt and Jane Luu "report the discovery of a very faint object with very slow . . . retrograde near opposition motion;" And: 1992 QB$_1$, "is currently between 37 and 59 AU from the earth;" And: some possible solutions for its still unknown orbit "are compatible with membership in the supposed Kuiper Belt"Brian G. Marsden, 1992, "1992 QB$_1$, IAU Central Bureau for Astronomical Telegrams," *Circular* No. 5611 (September 14, 1992).

—Jewitt, "We think this is the first of a large number of similar objects waiting to be discovered in the outer solar system.": John Noble Wilford ,1992, "Red Object Sighted Beyond Pluto May Be Part of Minor Planet Belt," *New York Times* (September 16, 1992).

—Jewitt, "By studying this and similar objects we can learn about the way the planets formed.": Kathy Sawyer, 1992, "Planet-Like Object Seen Beyond Pluto May Provide Clues to Origin of Comets," *Washington Post* (September 16, 1992).

—Kuiper proposed what he called "the belt just outside proto-Neptune, i.e., 38 to 50 AU from the Sun.": Gerard P. Kuiper, 1951, "On the Origin of the Solar System," in J. Allen Hynek, ed., *Astrophysics*, New York: McGraw-Hill Book Co., pp. 357–424.

—"Kuiper's 1951 paper was beaten to the punch by Edgeworth's 1949 . . . paper, yet Edgy seems to have been neglected by history.": David Jewitt, 1992, private communication (e-mail) to Maran (September 16, 1992).

—*Science* magazine called the discovery of 1992 QB$_1$ "a post-humous triumph" for Kuiper, and compared the importance of the event with the discovery of the first asteroid in 1801.: Richard A. Kerr, 1992, "Planetesimal Found Beyond Neptune," *Science*, 257, p. 1865.

—"This hypothesis implies that the present population of planets in the outer Solar System is much larger than previously recognized.": Alan Stern, 1991, "On the Number of Planets in the Outer Solar System: Evidence of a Substantial Population of 1000-km Bodies," *Icarus*, 90, p. 271–281.

—"a reservoir of icy, comet-size objects on the fringes of the solar system.": John Noble Wilford, 1995, "Scientists Find Source of Comets on Outer Edges of the Solar System," *New York Times* (June 15, 1995).

—Levison asserted that the seeming discovery made the Kuiper Belt "the most populous region in the planetary system.": John Noble Wilford, 1995, *ibid*.

—*Texas Monthly* described Cochran as a person who "talks about her work with the full-tilt enthusiasm that other people reserve for sports teams.": Helen Thorpe, 1995, "The Outer Limits," *Texas Monthly* (August 1995).

—"when astronomers don't find what they are looking for, the defeat can provide as much information as a successful search." And: "The outer solar system . . . appears not as crowded as some theories suggest, perhaps because small KBOs have already stuck together to form larger bodies or frequent collisions have ground down small KBOs into even smaller bits below the threshold of the survey.": Harvard-Smithsonian Center for Astrophysics Press Release 2008-18, "Outer Solar System Not as Crowded as Astronomers Thought" (October 3, 2008).

—ULTRACAM, described as a "high-speed triple-beam imaging photometer.": Francoise Roques, et al. 2006, "Exploration of the Kuiper Belt by High-Precision Photometric Stellar Occultations: First Results," *Astronomical Journal*, 132, 819–822.

—Varuna's "surface is darker than Pluto's, suggesting that it is largely devoid of fresh ice, but brighter than previously assumed for KBOs.": David Jewitt, Hervé Aussel, and Aaron Evans 2001, *Nature*, 411, 446–447.

—"Varuna may be a rotationally distorted rubble pile, with a weak internal constitution due to fracturing by past impacts.": David Jewitt and Scott S. Sheppard, 2002, "Physical Properties of Trans-Neptunian Object (20000) Varuna," *Astronomical Journal*, 123, pp. 2110–2120.

—Sedna, in an orbit "unexpected in our current understanding of the solar system.": Michael E. Brown, Chadwick Trujillo, and David Rabinowitz, 2004, "Discovery of a Candidate Inner Oort Cloud Planetoid," *Astrophysical Journal*, 617, pp. 645–649.

—Eris "stirred up a great deal of trouble among the international astronomical community when the question of its proper designation led to a raucous meeting of the IAU in Prague.": http://web.gps.caltech.edu/~mbrown/planetlila/

—"Eris has a mass 27% higher that that of Pluto (with an uncertainty of only 2%).": http://web.gps.caltech.edu/~mbrown/planetlila/moon/index.html

—Brown wrote to the Minor Planet Center on August 15, 2005, requesting that the Spaniards "be stripped of the official discovery status" and that the IAU also "issue a statement condemning their actions.": Dennis Overbye, 2005, "One Find, Two Astronomers: An Ethical Brawl," *New York Times* (September 13, 2005).

—Sedna as a "candidate inner Oort cloud planetoid.": Michael E. Brown, Chadwick Trujillo, and David Rabinowitz, 2004, *loc. cit.*

Chapter 10: Exoplanet Wars

A number of excellent books have been written on the recent discoveries of exoplanets. The most up-to-date, at the time of this writing, is *The Crowded Universe*, by Alan Boss (New York: Basic Books, 2009). We can also recommend *New Worlds in the Cosmos: The Discovery of Exoplanets* by two of the pioneers in the field, Michel Mayor and Pierre-Yves Frei (Cambridge: Cambridge University Press, 2003).

—"If an astronomer wishes to make front-page news, the surest way for him to proceed is to claim the discovery of a major planet.": Brian G. Marsden, 1980, "Planets and Satellites Galore," *Icarus*, 44, 29–37.

—The *New York Times* trumpeted "Another Solar System Is Found 36 Trillion Miles From the Sun.": Anonymous, 1963, *New York Times* (April 19, 1963).

—Burton called van de Kamp "the master" of astrometry, which, Burton wrote, "requires that its practitioners be meticulous, patient and persistent.": W. Butler Burton, 1986, "Introduction," in Peter van de Kamp, ed., *Dark Companions of Stars*, Dordrecht: D. Reidel Publishing Company, pp. 221–223 (reprinted from *Space Science Reviews*, 43, Nos. 3/4).

—Van de Kamp once quoted the *Talmud*, "If you want to understand the invisible, look careful at the visible.": Peter van de Kamp, 1986, "Preface," in Peter van de Kamp, ed., *Dark Companions of Stars*, Dordrecht: D. Reidel Publishing Company, pp. 219–220 (reprinted from *Space Science Reviews*, 43, Nos. 3/4).

—VB 8B was "what may be the first planet ever observed outside the solar system" and "If the discovery is verified, it would climax a centuries-old quest to find such a body.": National Science Foundation, "Astronomers Report Discovering First Planet Outside Solar System," Press Release NSF PR84-73 (December 11, 1984).

—The *New York Times*, which had been tipped off to the impending announcement, ran a story on the front page on the morning of December 11, headlined "Possible Planet Found Beyond Solar System.": John Noble Wilford, 1984, "Possible Planet Found Beyond Solar System," *New York Times* (December 11, 1984).

—VB 8B "is a Brown Dwarf and NOT a planetary object,": U.S. Naval Observatory 1984, "USNO Astronomers Say New Object is Not a Planet," Press Release (December 11, 1984).

—VB 8B was an astounding case of an astronomical object independently "discovered" by two groups, each of which was mistaken. *Sky & Telescope* magazine speculated that the errors may have come from "subtle systematic effects of Earth's atmosphere on the observations.": Ronald A. Schorn, 1987, *Sky & Telescope*, 74, VB 8B's Vanishing Act, p. 139.

—Bruce Campbell, "I think this is the best evidence of other planetary systems to date.": Deborah Blum, 1987, "More solar systems in galaxy?" *Sacramento Bee*, p. A1 (June 19, 1987).

—The director of planetary research at Lowell Observatory stated that the method devised to measure the wobbles "appears to be a major step forward.": Anonymous (distributed by Associated Press), "Evidence indicates other planets exist," Fort Lauderdale *Sun-Sentinel* (June 19, 1987).

—"there is strong evidence" that "we have observed a planet" of about 10 times the mass of Earth or somewhat more, "in a six-month circular orbit with a radius of 0.7 AU.": Matthew Bailes, Andrew G. Lyne, and Setnam L. Shemar, 1991, "A planet orbiting the neutron star PSR1829-10," *Nature*, 352, pp. 311–313.

—"The history of the searches for other planetary systems is littered with published detections that vanish under further scrutiny.": David Black, 1991, "It's all in the timing," *Nature*, 352, pp. 278–279.

—Lyne "just sat there frozen to his chair for the next half hour as the enormity of the mistake sunk in." And: "He could have simply published a retraction, but in a display of extraordinary scientific courage and honesty Lyne decided to announce the error" at the meeting.: Geoff McNamara, 2008, *Clocks in the Sky: The Story of Pulsars*, Berlin: Springer, p.143.

—Lyne's "honest retraction highlighted the scientific process at its finest.": Geoffrey Marcy, 1998, "Back in focus," *Nature*, 391, p. 127.

—detection of the three companions of PSR B1257+12 "constitutes irrefutable evidence that the first planetary system around a star other than the sun has been identified.": Alexander Wolszczan, 1994, "Confirmation of Earth-Mass Planets Orbiting the Millisecond Pulsar PSR B1257+12," *Science*, 264, pp. 538–542.

—"The announcement was greeted by heavy applause and some doubt." And: Franco Pacini, "How long could a planet last so close to the principal star without evaporating from the effect of the enormous quantity of energy it absorbed?": Daniel Williams, 1995, "Italian Astronomers Say a Jupiter-like Planet Circles a Star in Pegasus," *Washington Post*, p. A36 (October 8, 1995).

—"The jury is still out, but at this time I would bet on 51 Pegasi B being a brown dwarf, not a member of a planetary system.": David Black, 1996, "Other Suns, Other Planets?" *Sky & Telescope*, 92, No. 2, pp. 20–27.

—"As the presence of a planet will not influence the shapes of spectral lines, these variations are likely to reflect a hitherto unknown mode of stellar oscillation. The presence of a planet is not required to explain the data.": David F. Gray, 1997, "Absence of a planetary signature in the spectra of the star 51 Pegasi," *Nature*, 385, pp. 795–796.

—"Could the first planet discovered around a sunlike star be a mirage?": Corey S. Powell, 1997, *Scientific American* (May 1997), pp. 17–18.

—*Science* titled its story, "Is First Extrasolar Planet a Lost World?": James Glanz, 1997, "Is First Extrasolar Planet a Lost World?," *Science*, 275, pp. 1257–1258.

—Gray, "I've been amazed at the excitement, publicity and emotion this issue has raised.": Alexandra Witze, 1997, "Debate rages over existence of far-off planet," *Dallas Morning News* (August 4, 1997).

—January 1998, Gray published new findings in *Nature*. He conceded that "a planet may indeed by the best explanation": David F. Gray, 1998, "A planetary companion for 51 Pegasi implied by absence of pulsations in the stellar spectra," *Nature*, 391, pp. 153–154.

—Marcy wrote that those astronomers like himself who were upset with Gray's first paper, "may occasionally forget that competition and human emotion have always provided fuel for the vigorous pursuit of alternative theories. It was right that the planet interpretation should not go unchallenged.": Geoffrey Marcy, 1998, *loc. cit.*

—Senator Jake Garn said after a ride in orbit aboard the Space Shuttle, "I gazed out into space, and it was clear to me that there had to be life out there.": Howard Blum, *New York Times Magazine, SETI Phone Home* (October 21, 1990).

CHAPTER 11: WHAT'S NEXT FOR PLUTO

The book on the future of Pluto has not yet been written. But when it is written, it will surely rely heavily on the results of the New Horizons mission. You can follow New Horizons at its websites: http://pluto.jhuapl.edu/ and http://www.nasa.gov/mission_pages/new horizons/main/index.html. For other opinions and anecdotes on Pluto, we refer readers to Neil deGrasse Tyson's *The Pluto Files*, and Govert Schilling's *The Hunt for Planet X*, mentioned several times earlier in these notes.

—"The facts or truths we discover in science come from thinking about the problem and developing a consensus through the scientific method that guides our research. Perhaps the lack of consensus really tells us that we don't know enough yet to define what a planet is." http://www.boulder.swri.edu/~buie/pluto/iauresponse.html

—"a messy problem was cleaned up by the creation of a new category of objects called 'dwarf planets.'" William F. Bottke, 2007, "Is Pluto a Planet?," book review in *Physics Today*, 60, No. 10, p. 55.

RECOMMENDED READING

SELECTED READINGS ON PLUTO, THE KUIPER BELT, AND THE HISTORY OF PLANET AND EXO-PLANET DISCOVERY

Baum, Richard, and William Sheehan. *In Search of Planet Vulcan: The Ghost in Newton's Universe.* New York: Plenum Trade, 1997.

Boss, Alan. *The Crowded Universe: The Search for Living Planets.* New York: Basic Books, 2009.

Davies, John. *Beyond Pluto.* Cambridge: Cambridge University Press, 2001.

Grosser, Morton. *The Discovery of Neptune.* Cambridge: Harvard University Press, 1962.

Hoskin, Michael. *The Herschel Partnership.* Cambridge: Science History Publications, 2003.

Hoyt, William Graves. *Planets X and Pluto.* Tucson: University of Arizona Press, 1980.

Jewitt, David, Alessandro Morbidelli, and Heike Rauer. *Trans-Neptunian Objects and Comets.* Berlin: Springer, 2008.

Lemonick, Michael. *The Georgian Star: How William and Caroline Herschel Revolutionized Our Understanding of the Cosmos.* New York: W. W. Norton & Company, 2008.

Littmann, Mark. *Planets Beyond: Discovering the Outer Solar System.* Mineola, NY: Dover Publications, 2004.

Lubbock, Constance, ed. *The Herschel Chronicle: The Life-Story of William Herschel and His Sister Caroline Herschel.* Cambridge: Cambridge University Press, 1933.

Maran, Stephen P., and Laurence A. Marschall. *Galileo's New Universe: The Revolution in Our Understanding of the Cosmos.* Dallas, TX: BenBella Books, 2009.

Mayor, Michel, and Pierre-Yves Frei. *New Worlds in the Cosmos: The Discovery of Exoplanets.* Cambridge: Cambridge University Press, 2003.

Putnam, William Lowell. *The Explorers of Mars Hill: A Centennial History of Lowell Observatory, 1894-1994.* West Kennebunk, ME: Phoenix Publishing, 1994.

Roseveare, N. T. *Mercury's Perihelion from LeVerrier to Einstein.* Oxford: Clarendon Press, 1982.

Schilling, Govert. *The Hunt for Planet X: New Worlds and the Fate of Pluto.* New York: Copernicus Books, 2009.

Sheehan, William. *The Planet Mars: A History of Observation and Discovery.* Tucson: University of Arizona Press, 1996.

Sheehan, William. *Worlds in the Sky: Planetary Discovery from Earliest Times Through Voyager and Magellan.* Tucson: University of Arizona Press, 1992.

Stern, Alan, and Jacqueline Mitton. *Pluto and Charon: Icy Worlds on the Ragged Edge of the Solar System.* New York: John Wiley and Sons, 1998.

Tyson, Neil deGrasse. *The Pluto Files: The Rise and Fall of America's Favorite Planet.* New York: W. W. Norton & Company, 2009.

Weintraub, David A. *Is Pluto a Planet? A Historical Journey through the Solar System.* Princeton: Princeton University Press, 2007.

PLUTO ON THE WEB

An extensive Pluto bibliography by Robert Marcialis, Lunar & Planetary Laboratory, University of Arizona: http://www.lpl.arizona.edu/~umpire/science/plubib_home.html.

The Great Planet Debate: Watch a video of scientists debating the status of Pluto at a debate held at Johns Hopkins University in August 2008: http://gpd.jhuapl.edu/.

IAU Web pages on the Pluto Resolution: http://www.iau.org/public_press/news/release/iau0603/.

Planets that weren't: http://www.solarviews.com/eng/hypothet.htm.

The Society for the Preservation of Pluto as a Planet: http://www.plutoisaplanet.org/.

ABOUT THE AUTHORS

Dr. Stephen P. Maran spent more than 35 years in NASA, working on the Hubble Space Telescope and other scientific projects and is the press officer for the American Astronomical Society. His 11 previous books include *Astronomy for Dummies*® and *The Astronomy and Astrophysics Encyclopedia*. His awards and honors include the naming of an asteroid for him by the International Astronomical Union, the NASA Medal for Exceptional Achievement, the George Van Biesbroeck Prize of the American Astronomical Society and the Astronomical Society of the Pacific's Klumpke-Roberts Award for outstanding contributions to the public understanding and appreciation of astronomy.

Laurence Marschall, PhD, is the W.K.T. Sahm Professor of Physics at Gettysburg College where he teaches courses in astronomy, physics and science writing. He writes a regular column on science books of note for *Natural History* magazine and serves as deputy press officer of the American Astronomical Society. In addition to more than 40 articles in professional journals, Marschall has written for publications such as *Sky and Telescope*, *Astronomy*, *Natural History*, *Discover*, *Harper's*, *Newsday* and *The New York Times Book Review*. His book *The Supernova Story* (Princeton Science Library, 1994) has been widely praised for its readability. He was awarded the 2005 Education Prize of the American Astronomical Society for his work in furthering undergraduate instruction in astronomy.

Maran and Marschall are also authors of *Galileo's New Universe* (BenBella Books, 2009).